Annual Conference of the American Forage and Grassland Council (AFGC 2024)

Mobile, Alabama, USA
7 - 10 January 2024

ISBN: 978-1-7138-9928-0

Printed from e-media with permission by:

Curran Associates, Inc.
57 Morehouse Lane
Red Hook, NY 12571

Printed by Curran Associates, Inc. (2025)

American Forage and Grassland Council (AFGC)
PO Box 867
Berea, KY 40403
USA

Phone: 800.944.2342

www.afgc.org

Additional copies of this publication are available from:

Curran Associates, Inc.
57 Morehouse Lane
Red Hook, NY 12571 USA
Phone: 845-758-0400
Fax: 845-758-2633
Email: curran@proceedings.com
Web: www.proceedings.com

TABLE OF CONTENTS

AGRONOMIC AND NUTRITIVE VALUE RESPONSES OF WARM-SEASON ANNUAL FORAGE MIXTURES MANAGED UNDER GRAZING

L.S. Silva, K. Seavey, D. Adkison[1]

Annual warm-season forages are fast-growing, high-quality options that complement forage production and support animal performance in livestock operations. In the Southeast, they are planted from April through June and can be managed for baleage production or under grazing. This study was conducted to determine forage production and quality, ground coverage and root biomass of three warm-season forage mixtures managed under rotational stocking. The study was conducted at the Edisto Research and Education Center in Blackville, South Carolina. Three treatments consisting of three forage mixtures, Grass [pear-millet (PM) + crabgrass (CG)], Mix1 [PM + sunnhemp (SH) + cowpea + soybean (SOY)], and Mix2 [PM + CG + SH + SOY], were arranged on a randomized complete block design and replicated three times. Grasses ranged from 65 to 80% of the total botanical composition of the mixtures, while the legume-containing treatments averaged 20% of legumes throughout the season. Forage accumulation per grazing cycle ranged from 2572 to 3330 lb/A for forage mixtures. Neutral and acid detergent fiber concentrations ranged from 57 to 60% and 31 to 35%, respectively. Root biomass for the grass-only treatments was approximately 20% inferior to other treatments. Based on the results, warm-season forage mixtures are a viable alternative to complement forage production and quality to support animal performance, while adding organic material into soil during summer.

[1]Assistant Professor, Clemson University, Blackville, SC (lseveri@clemson.edu); Master Student, Clemson University, Blackville, SC (kseavey@clemson.edu); Field Technician, Clemson University, Blackville, SC (dadkiso@clemson.edu).

SEED PRODUCTION COMPONENTS OF NEW TETRAPLOID BAHIAGRASS LINES FROM THE UNIVERSITY OF FLORIDA FORAGE BREEDING PROGRAM

R.P. Becker, E.F. Rios, K.E. Kenworthy, L.E. Sollenberger, M. Wallau, and N. Caram[1]

Abstract

Bahiagrass (*Paspalum notatum* Flügge) is a C4 perennial warm-season grass largely used in the southern USA by beef cattle producers. Challenges to the establishment of new areas are still present because of low seed production and quality characteristics. This study aimed to evaluate the seed production of 9 new tetraploid bahiagrass lines compared with Argentine under two cutting heights (5 and 10 cm). The experiment was carried out at the Beef Research Unit of the University of Florida, Gainesville, FL, using a four replicates of a randomized complete block design with a 10 x 2 factorial treatment arrangement (10 entries x two cutting heights) during summer and fall 2022 and 2023. Data were collected in each subplot, and the variables measured were seed yield, filled seed production, number of seed heads/m^2, seed set and thousand seed weight. Entries were significatively different for all the variables measured ($\alpha<0.05$). 3Fpen8 had the highest yield in both years followed by Hybrid 3 in 2022 and by M34, FPN1901 and M27 in 2023 (Table 1). The filled seed production in 2022 was greater for 3Fpen8 and Hybrid 3 and in 2023 only for 3Fpen8 (Table 1). Seed heads was affected by entries, years and cutting heights. 3Fpen8 and Hybrid 3 had the highest number of seed heads/m^2 for both cutting heights (Table 2). The lower cutting height (5 cm) increased seed heads/m^2 for Hybrid 3, M27, M34, and M98Alt entries. 3Fpen8 and Hybrid 3 reduced seed heads/m^2 and Argentine and M6Alt increased seed heads/m^2 in 2023 vs. 2022. This reduction matches with the reduction in seed yield for 3Fpen8 and Hybrid 3 in 2023, while Argentine and M6Alt increased both seed heads/m^2 and seed yield in 2023. Seed set was affected by entry and year ($P<0.031$ and <0.0001, respectively). There was also an entry*year interaction with M27 having the highest seed set in 2022 (15.5%). In 2023, 3Fpen8 had the highest seed set (15.4%). In the second, M98Alt, Argentine, A4EMS, M27 and Hybrid 3 decreased and 3Fpen8 increased seed set (Table 3). Thousand seed weight was also affected by entry and year with 3Fpen8 having the highest value in 2022 (3.64 g) and 3Fpen8, M34, FPN1901 and Hybrid 93 having the highest values in 2023. There was an interaction year*entry ($P=0.035$), with M27, 3Fpen8, M34, FPN1901, A4EMS and Hybrid 3 having lesser thousand seed weight in the second year (Table 4). Lower rainfall and probable metsulfuron herbicide damage may explain some of the differences between years and entries related to the variables. 3Fpen8 is the entry with the best performance in seed yield and production of filled seeds, outyielding Argentine in both years of assessment.

Introduction: Bahiagrass is a C4, perennial warm-season grass very adapted forage species and broadly used by beef cattle producers on the Southeast USA due to his characteristics of low fertility requirement, adaptation to the region climate conditions and grazing tolerance (Wallau *et al.*, 2019; Chambliss and Sollenberger, 1991). The main cultivars available in the USA are diploids as the hybridization of tetraploid is more difficult related to their apomitic reproduction pathway.

[1]University of Florida-IFAS Agronomy Department, Gainesville, FL 32608.

According to Adkins *et al.*, 2002, warm season grasses still have several wild species characteristics related to seed dormancy, a key qualitative characteristic to improve pasture establishment. Low seed production, low seed quality and subsequent low germination rates are some of the challenges found in bahiagrass (Rios *et al.*, 2020). Relevant studies suggest that seed productivity is mainly improved by higher seed head density which is stimulated by tillering. Also, practices such as canopy renewing in spring followed by intermediate levels of nitrogen fertilization appeared to be good approaches to improve seed production (Rios *et al.*, 2020; Adjei *et al.*, 2000). The objective of this work is to assess effect of spring cutting heights on tillering and seed production of new tetraploid bahigrass lines.

Materials and Methods: The experiment was carried out at the Beef Research Unit of the University of Florida, Gainesville, Florida during Spring to Summer of 2022 and 2023. The soil is classified as Chipley sand. The experiment was a randomized complete block in a factorial 10 x 2 with 4 replications. One factorial was the bahiagrass entries (3Fpen8; A4EMS; M6Alt; M34; FPN1901; M98Alt; M27; Hybrid 3; Hybrid 93; Argentine); and the other was the two spring cutting heights (5 and 10 cm above the soil surface). The bahiagrass entries were established during summer 2019 in 3 m x 3 m plots, managed under mob-stocking during summer and autumn of 2020 and 2021.

All the plots were split in half and mowed on May 23rd, 2022, and June 19th, 2023, using a flail collection mower Peruzzo 1200. After the cuttings the plots were fertilized with 54 lb N/acre, rate that increases the bahiagrass seed production (Rios, 2020). In May 2023, before mowing, there was a metsulfuron application to control diploid type bahiagrass. Seed yield components were measured at each subplot on each harvest day within a 0.25-m^2 quadrat placed in the center of each experimental unit. Seedheads were counted, harvested and the number extrapolated to seed heads/m^2. After the harvest, all the seed heads were dried in an air forced dryer at 35°C and then threshed. The seed yield was the summation of filled and empty florets right after the threshing. All the harvested material was passed through the forced-air seed blower (757 South Dakota Seed Blower, Seedburo Equipment) to separate the empty florets and other inert material. Extrapolations were done to give the yield and filled seeds produced in lb/acre. We assessed the seed weight by counting 3 samples of 100 filled florets for each filled and empty florets. The thousand seed weight was calculated as the average of the three 100 seeds sample extrapolated to one thousand seeds. The seed set was calculated as the ratio between the number of filled florets to the total number of florets produced. The statistical analysis was done using SAS where the least squares means were compared by simple LSD (α =0.05).

Results and Discussion: All the variables analyzed were significatively different between entries and had a year interaction (P<0.05). In 2022, yield was higher for 3Fpen8 and Hybrid 3 (768 and 680 lb/acre, respectively). In 2023, Hybrid 3 reduced seed yield and the highest values were produced by 3Fpen8, M34, FPN1901 and M27 (Table 1). 3Fpen8, Hybrid 3 and A4EMS produced less yield in 2023, on the other way M6Alt and Argentine increased yield in 2023 (Table 1). The filled seed production in 2022 was higher for 3Fpen8 and Hybrid 3 and in 2023 only for 3Fpen8 (Table 1). Hybrid 3, M27, Hybrid 93 and Argentine produced less filled seed in 2023 (Table 1). These results might be explained by lower rainfall in 2023 summer.

The number of seed heads/m^2 was affected by height, entry, and had height*entry and year*entry interactions. 3Fpen8 and Hybrid 3 had the highest numbers on both cutting heights

(Table 2). This result infers the importance of seed heads density as an important seed yield component. The lower cutting height (5 cm) increased seed heads/m^2 for Hybrid 3, M27, M34, and M98Alt entries. When comparing years, 3Fpen8 and Hybrid 3 reduced seed heads/m^2 and Argentine and M6Alt increased seed heads/m^2 in 2023 (Table 2). This reduction matches with reduction on seed yield for 3Fpen8 and Hybrid 3 in 2023. On the other hand, Argentine and M6Alt increased both seed heads/m^2 and seed yield in 2023. A third situation also must be pointed out, A4EMS did not decrease seed heads/m^2 in the second year but reduced yield, showing that other limitations or driver components might be involved in seed yield.

Seed set was affected by entry and year ($P<0.031$ and <0.0001, respectively). There was also an entry*year interaction ($P<0.0017$). M27 had the highest seed set in 2022 (15.5%). In 2023 3Fpen8 had the highest (15.4%). In the second year M98Alt, Argentine, A4EMS, M27 and Hybrid 3 decreased the Seed set and 3Fpen8 increased (Table 3).

Thousand seed weight was also affected by entry and year ($P<0.0001$ and 0.0272, respectively). 3Fpen8 had the highest value in 2022 (3.64 g). In 2023, 3Fpen8, M34, FPN1901 and Hybrid 93 had the highest values. There was an interaction year*entry ($P<0.0351$). M27, 3Fpen8, M34, FPN1901, A4EMS and Hybrid 3 had a reduction in thousand seed weight in the second year (Table 4).

The period comprised from May to August 2022 had a 24 in of rainfall in which July and August months (flowering period for bahiagrass) had over 7 inches each month. In 2023, the same period has 22 in rainfall, however, the July and August months had 5 and 4 in rainfall, respectively, which may have reduced the water availability to the plants and restricted yield and filled seeds production to some entries. Although some entries were able to increase seed yield in the second year, there was not the same response when analyzing amount of filled seeds produced by the entries. This suggests that the ability of the entries in dealing with stresses are different. 3Fpen8 and A4EMS produce less yield but can keep the same filled seed production. On the other hand, Argentine and M6Alt increased seed yield but for M6Alt it did not reflect in higher filled seed production and for the Argentine it was lower in the second year.

Seed heads density is one of the most important seed yield components affected by management. Argentine and M6Alt increased seed heads/m^2 in the 2023, same year were the seed yield increased for both. 3Fpen8 and Hybrid 3 decreased seed heads/m^2 in 2023 and seed yield decreased for both. For A4EMS the yield was also lower in 2023 but the seed heads/m^2 was not different. Argentine and M6Alt were able to increase the seed heads density in the second year which were able to produce higher seed yield, but it was not converted it into higher filled seeds production.

Another factor that may have affected the variables was the metsulfuron herbicide application done on beginning of May 2023 to control diploid bahiagrass infestation on the plots. Bunnell *et al.*, 2003, showed that tetraploid bahiagrass types have some growth reduction when metsulfuron is applied. This aspect besides the lower rainfall may have impacted the responses of the entries.

Literature cited:

Adjei, M.B., Mislevy, P., Chason, W. 2000. Timing, Defoliation Management, and Nitrogen Effects on Seed Yield of 'Argentine' Bahiagrass. Agronomy Journal, v. 92, n. 1, p. 36-41. doi: https://doi.org/10.2134/agronj2000.92136x.

Adkins, S.W., Bellairs, S.M., & Loch, D.S. 2002. Seed dormancy mechanisms in warm season grass species. Euphytica, 126, 3-20. doi: https://doi.org/10.1023/A:1019623706427

Bunnell, B. T. Bunnell, B. T., R. D. Baker, L. B. McCarty, D. W. Hall, and D. L. Colvin. 2003. Differential response of five bahiagrass (Paspalum notatum) cultivars to metsulfuron. Weed Technol. 17:550–553, doi: https://doi.org/10.1614/0890-037X(2003)017[0550: DROFBP] 2.0. CO;2

Chambliss, C.G., and Sollenberger, L.E. 1991. "Bahiagrass: The foundation of cow-calf nutrition in Florida." Proceedings of the 40th Annual Beef Cattle Short Course. Gainesville, Florida: IFAS, University of Florida, Gainesville, Florida: 74-80.

Rios, E.F. *et al.* 2020. Managing forage and turf-type bahiagrass for seed production. Crop Science, v. 60, n. 3, p. 1569-1579. doi: https://doi.org/10.1002/csc2.20146

Wallau, M., Vendramini, J., Dubeux, J., and Blount, A. 2019. Bahiagrass (Paspalum notatum Flueggé): Overview and Pasture Management. University of Florida IFAS Extension. doi: https://doi.org/10.32473/edis-ag342-2019

Table 1. Seed yield and filled seed production of the 10 tetraploid bahiagrass lines on the two years of assessment.

	Entry									
Year	3Fpen8	Hyb3	M34	FPN1901	M27	A4EMS	Hyb93	Argentine	M98Alt	M6Alt
Yield	lb/acre									
2022	768 Aa	680 Aa	283 Ab	259Ab	231Ab	215Abc	136Ac	70Bd	49Ad	18Be
2023	319 Ba	143 Bbc	207 Aab	200Aab	202Aab	131Bbc	103Ac	134Abc	42Ad	54Ad
Filled seed production										
2022	151Aa	122Aa	67Ab	61Abc	62Abc	38Acde	38Ad	25Ade	21Aef	9Af
2023	113Aa	16Bbcde	46Ab	41Ab	33Bbc	14Acde	18Bcde	17Bcd	6Ade	4Ae

[a]Means followed by different uppercase letters are different by LSD values ($P<0.05$) within a given entry.
[b]Means followed by different lowercase letters are different by LSD values ($P<0.05$) within a given year.

Table 2. Number of seed heads/m^2 the 10 tetraploid bahiagrass lines at each cutting height.

	Entry									
Height	3Fpen8	Hyb 3	FPN1901	A4EMS	M27	M34	Argentine	Hyb93	M6Alt	M98Alt
(cm)	Seed heads/m^2									
5	359Aa	269Aa	131Ab	121Abc	161Ab	171Ab	76Acd	147Abc	54Ade	35Ae
10	251Aa	197Ba	136 Ab	120Abc	101Bc	91 Bbc	77Ac	55Acd	34Ad	21Bd
year										
2022	400Aa	348Aa	144Ab	133Ab	125Ab	132Ab	47Bc	108Ab	23Bcd	22Ad
2023	210Ba	118Babc	123Ab	108Abc	137Aab	129Aab	106Abc	94 Abc	65Ac	34Ad

[a]Means followed by different uppercase letters are different by LSD values ($P<0.05$) within a given entry or year.
[b]Means followed by different lowercase letters are different by LSD values ($P<0.05$) within a given height.

Table 3. Seed set of the 10 tetraploid bahiagrass lines on the two years.

	Entry									
	M98Alt	Argentine	A4EMS	M34	M27	Hyb 93	M6Alt	FPN1901	3Fpen8	Hyb 3
Year	%									
2022	26.2Aab	21.0Aab	19.4Aab	15.8Aab	15.5Aa	15.5Aab	15.4Aab	13.9Aab	11.8Bab	9.8Ab
2023	7.8Bbcd	5.5Bcd	4.6Bde	12.9Aab	8.3Bb	9.2Abc	3.7Ae	10.8Aab	15.4Aa	4.4Bde

[a]Means followed by different uppercase letters are different by LSD values ($P<0.05$) within a given entry.
[b]Means followed by different lowercase letters are different by LSD values ($P<0.05$) within a given year.

Table 4. Thousand seed weight of the 10 tetraploid bahiagrass lines on the two years.

	Entry									
	M27	3Fpen8	M34	FPN1901	Hyb 93	A4EMS	Hyb 3	Argentine	M98Alt	M6Alt
Year	g									
2022	3.65Aa	3.64Aa	3.45Aab	3.26Abc	3.12Acd	3.08Abcd	3.01Ad	2.773Abcd	2.55Acd	2.47Abcd
2023	3.01Ba	3.09Ba	3.05Ba	2.95Ba	3.01Aa	2.44Bb	2.43Bb	3.02Aab	2.77Aab	2.80Aab

[a]Means followed by different uppercase letters are different by LSD values ($P<0.05$) within a given entry.
[b]Means followed by different lowercase letters are different by LSD values ($P<0.05$) within a given year.

STRATEGIES TO OVERSEED FORAGE LEGUMES ON LIMPOGRASS IN SUMMER

Marilia Bernardini[1], Jose C.B. Dubeux Jr. [1], Kenneth Oduor[1], Kevin Roger Trumpp[1], Javier Portuguez[1], Igor Bretas[1], Luana D. Queiroz[1], and M.Ruiz-Moreno

Limpograss (*Hemarthria altissima*) is a warm-season grass and has been successfully adopted in South Florida. Limpograss is not as light sensitive as bahiagrass or bermudagrass, and it is one of the first warm season grasses to regrow after a frost. Stockpiled limpograss has a reasonable digestibility, but its CP concentration is low. Forage legumes could be a potential option to improve crude protein. Forage legumes can also add nitrogen to the system through biological nitrogen fixation. This project was conducted at UF IFAS North Florida Research and Education Center in Marianna, FL. Treatments included 2 legume species (sunn hemp and aeschynomene) and three seeding rates (8, 16, and 24 lb/A), and two controls (limpograss + 50 lb N/A and unfertilized limpograss). Aeschynomene, sunn hemp, and limpograss are morphologically different and samples were hand-separated after each harvest. Forage accumulation of each component was calculated as the product of the forage accumulation and the proportion of the component (limpograss, legumes). Limpograss herbage mass ranged from 3840 to 9956 kg/ha and averaged 6900 kg DM/ha. Legume herbage mass did not contribute much for the total herbage mass. Expected results from this research is the recommendation of the best combination of summer legume and seeding rate to integrate in 'Gibtuck' limpograss swards.

[1]University of Florida, North Florida Research and Education Center-Marianna, 3925 Highway 71, Marianna, FL, 3925 USA

INCREASING FORAGE PLANT FAMILY DIVERSITY AND SUBSEQUENT IMPACTS ON CROP PRODUCTION

Eric D. Billman[1]

As integrated crop-livestock systems become more prevalent, understanding the impact that annual forage production may have on subsequent row crop growth and yields is vital to optimizing yields of both forages and cash crops. This study compared the effects of diversifying forage plant families on a cotton crop in South Carolina. Monocultures of cereal rye, crimson clover, and carinata (*Brassica carinata* A. Braun), along with all possible two- and three-way mixtures were planted in fall 2022, then terminated and roller-crimped in April 2023. Cotton was subsequently strip-tilled into plots in early May. Forage species persistence and biomass were evaluated in the spring, while soil moisture, cotton moisture stress, plant morphology, soil fertility, and cotton yields were evaluated on the cotton crop throughout 2023. Results indicated that treatments with only crimson clover and carinata had 60 – 70 lb NO_3-N compared to 48 – 52 lb NO_3-N for any treatment that contained cereal rye. The three-species mixture reduced individual species persistence by 30 – 60%, but reduced weed pressure by >90% during spring months. Finally, cotton grown following any monoculture of mixture containing cereal rye and carinata had greater seed cotton yields than when following crimson clover. The increased nitrogen from crimson clover and carinata likely contributed to excess nitrogen availability in the growing cotton crop, resulting in taller plants that produced fewer bolls, thus reducing yields. This necessitates changes in cotton fertility management to reduce this risk when following a legume or brassica crop.

[1]Research Agronomist, USDA-ARS, Coastal Plains Soil, Water, and Plant Research Center, Florence, SC (eric.billman@usda.gov).

SOIL-TEST BIOLOGICAL ACTIVITY IN A GRASS AND GRASS-LEGUME GRAZING SYSTEMS

Beatriz E. Bizzuti Cremostim, José C.B. Dubeux, Flavia F. Simili, Luana M. D. Queiroz, Martin Ruiz-Moreno, Kevin R. Trumpp, Igor L. Bretas, and Mario de A. L. Junior[1]

The introduction of forage legumes in grazing systems may increase the livestock sustainability and improve the delivery of ecosystems services from grasslands. Grass-legume mixtures add N to the system via biological N2-fixation and might improve efficiency of N cycling. In addition, these mixtures might increase C deposition and improve litter quality, changing soil biological activity due to the plant diversity. The methodology flush of CO_2 has been used to see the soil biological activity, and has showed association with N mineralization. The aim of this investigation was to evaluate the soil biological activity, through the flush of CO_2, in three grazing systems including ´Argentine´ bahiagrass (*Paspalum notatum* Flugge) with N fertilizer application, grass with a blend of clovers (*Trifolium* spp.), and grass with clovers plus Rhizoma peanut (*Arachis glabrata* Benth.). There was no difference for CO_2 flush between the grazing systems. The treatments evaluated has similar biological activity, however the inclusion of legume forage in the grassland systems is important to reduce N fertilizer.

[1]Exchange Ph.D. student- Beatriz E. Bizzuti Cremostim (beatrize.bizzuti@ufl.edu), Professor - José C.B. Dubeux (dubeux@ufl.edu), Visiting Professor - Flavia F. Simili (similif@ufl.edu), Ph.D. student - Luana M. D. Queiroz (ldantasqueiroz@ufl.edu), Biological Scientist - Martin Ruiz-Moreno (ruizm001@ufl.edu), MS student - Kevin Roger Trumpp (ktrumpp@ufl.edu), Posdoctoral Associate - Igor L. Bretas (ig.limabretas@ufl.edu), Professor - Mario de A. L. Junior (lirajunior.ma@ufl.edu) – North Florida Research Education Center – Marianna, Florida.

WHY PURCHASE CERTIFIED SEED?

Ann Blount[1], Jim Bostick[2,3], Steven Hague[4], Liliane Severino da Silva[5], and Marnin Wolfe[6]

You know the old adage "you get what you pay for". That is the reasoning for purchasing certified seed. What exactly is the difference between "certified" and "non-certified seed" other than increased cost of seed? Many of us seldom look at the tag on the bag of seed or documentation of the vegetative plant materials that we purchase for our farming operations. Maybe a better understanding is needed of the role of this *internationally recognized* system that is used to identify crop purity/genetic assurance, which upholds its high quality and true-to-type integrity. The Association of Official Seed Certifying Agencies (AOSCA) was established in 1919 as the International Crop Improvement Association, whose certification membership includes agencies in the USA, Argentina, Australia, Brazil, Canada, Chile, New Zealand and South Africa. The purpose of AOSCA is the "preservation of the genetic purity and varietal identity", represented by "blue" (certified) seed tags that meet state, federal and international seed laws. AOSCA provides certification for some vegetatively-propagated crops as well, such as bermudagrass. So why purchase certified seed or planting materials? The strict standards for certified seed/vegetative stock guarantee that you will get a quality product, with high standards for germination and purity, minimal weed seed or off-types and legal traceability of the source of plant material. Today, many private seed/sod companies plant products, as well as public university plant variety releases, are protected through the AOSCA certification system.

[1]Germplasm Adv., Southern Seed Cert. Assoc., Inc., (greengenesgenetics@outlook.com); [2]Exec. Vice President, Southern Seed Cert. Assoc., Inc.; [3]Exec.Vice Pres., Alabama Crop Improvement (acia7383@gmail.com); and [4]Depart. Head, Crop, Soil and Env. Sci. Dept., Auburn Univ. (shague@auburn.edu); [5]Asst. Prof., Ext. Specialist, Animal and Vet. Sci. Dept., Clemson Univ. (lseveri@clemson.edu); [6]Asst. Prof., Quant. Genetics, Auburn Univ. (mdw0092@auburn.edu).

NUTRIENT REMOVAL COSTS OF TENNESSE HARVESTED FORAGES

C.N. Boyer[1], D. McIntosh[2], B.C. Pedreira[3], K. Mason[4], and C.C. Martinez[5]

Making hay for winter feeding is critically important for the profitability of cattle production in the Southeastern United States. Recent increases in fertilizer prices have concerned producers about their cost of hay production and the implications for profitability. To help producers ensure adequate fertilization and to price their hay, the nutrient removal costs were estimated in eight species of harvested forages in Tennessee. Forage sample data from 2015 to 2022 across (5,148 samples) Tennessee were utilized to determine nutrients removed and market prices to estimate the cost of nutrients removed. Findings conclude that orchardgrass removed the most total nitrogen (N), phosphorus (P), and potassium (K), which was 44 lb/1,000 bale of hay followed by annual rye (42 lb), warm season annuals (42 lb), and bermudagrass (39 lb). In terms of costs, N was the primary nutrient removed and contributed most of the cost, followed by K and finally P. On average, for a 1,000 lb bale of hay, the range of total nutrients removed was $20.07 for native warm-season grasses to orchardgrass having the highest cost of $26.82. A tall fescue mixed bale was estimated to have $24.19 in nutrients removed, which was slightly higher than tall fescue bale of $22.84. These estimated costs are helpful for producers to know their cost of feed and how to price their hay. This demonstrates how lab data can be utilized for economic analysis that are relevant for producer education in forage and livestock production.

[1]Professor in Department of Agricultural and Resource Economics (cboyer3@utk.edu); [2]Coordinator of UT Beef and Forage Center; [3]Associate Professor in Department of Plant Sciences and Director of UT Beef and Forage Center; [4]Assistant Professor in Department of Animal Science; and [5]Assistant Professor in Department of Agricultural and Resource Economics.

USING DRONE IMAGES AND ARTIFICIAL INTELLIGENCE FOR SITE-SPECIFIC WEED MANAGEMENT

Igor Bretas[1], Jose Dubeux[1], and Chang Zhao[2]

Weed encroachment negatively affects grassland productivity by reducing herbage allowance, stocking rates, and livestock performance. *Amaranthus spinosus* L. is an invasive weed species widely found in grasslands. The high costs of chemical application and the global concern about environmental impacts restrict indiscriminate herbicide spraying in grasslands. Site-specific weed management is a weed management strategy based on weed spot-spraying that has the potential to overcome these issues. Drone images and artificial intelligence can provide valuable information for weed mapping to drive herbicide application in grasslands. We developed a deep convolutional neural network (CNN) to detect and map *Amaranthus spinosus* in common bermudagrass [Cynodon dactylon (L.) Pers.] pastures using drone images. The study was carried out in twelve paddocks of approximately 1.3 ha each. The experiment was established in a randomized complete block design, with three treatments and 4 blocks. The treatments were bermudagrass paddocks free of weeds, paddocks with weeds established in alternated strips, and paddocks with weeds spread throughout the whole area. The images used for the model development were taken using a multirotor drone (model: DJI M100) equipped with an RGB camera (model: DJI X5). The CNN model was able to detect around 80% of the *Amaranthus spinosus* with an average prediction accuracy of 94%. Our weed mapping showed the potential of using the U-Net model to generate an herbicide application map to be inserted into the sprayer system, reducing up to 40% of the amount of herbicide applied in weed-infested areas.

[1]University of Florida, North Florida Research and Education Center-Marianna, 3925 Highway 71, Marianna, FL; [2] Agronomy Department, University of Florida, Gainesville, FL.

EVALUATING CHEMICAL CONTROL METHODS OF THE BERMUDAGRASS STEM MAGGOT

Justin C. Burt[1], Lisa L. Baxter[2], William F. Anderson[3], and Guy A. Hancock[4]

Bermudagrass stem maggot (BSM; *Atherigona reversura* Villeneuve) negatively impacts bermudagrass (*Cynodon spp.*) with respect to herbage accumulation (HA) and nutritive value. While all bermudagrasses are susceptible to BSM, *C. dactylon* cultivars are more vulnerable to the damage compared to those with *C. nlemfuensis* influence. Pyrethroid insecticides are essential to mitigate BSM but producers have considered the use of spinosad because of residual control of other insects or reducing pyrethroids to save money. Our objective was to compare the agronomic and economic impacts of one or more pyrethroid and/or spinosad applications on two bermudagrass cultivars. Eight insecticide treatments were evaluated on 'Alicia' and 'Tifton 85' bermudagrass. Data indicate that two applications of a pyrethroid resulted in the greatest herbage accumulation (HA) in both cultivars compared to the unsprayed control. The addition of spinosad did not improve HA compared to the unsprayed control. Crude protein and total digestible nutrients were not affected by insecticide treatment. Economic returns were improved with two pyrethroid applications compared to other insecticide treatments. The results of this study confirm that there is not yet an alternative for two pyrethroid applications for BSM suppression. Ongoing research is focusing on the improvement and development of more BSM tolerant cultivars to reduced reliance on the pyrethroid mode of action.

[1]Postdoctoral Researcher, Crop and Soil Science Department (justin.burt@uga.edu); University of Georgia Tifton Campus, Tifton, GA. [2]Assistant Professor and Extension Forage Specialist, Crop and Soil Science Department (baxterl@uga.edu); University of Georgia Tifton Campus, Tifton, GA. [3]Research Geneticist (bill.anderson@usda.gov), USDA-ARS, Tifton, GA. [4]Public Service Associate, Agricultural and Applied Economics Department (ghancock@uga.edu); University of Georgia Tifton Campus, Tifton, GA.

DEVELOPING A FORAGE GRASS MATURITY INDEX

Kim Cassida, Roeland Kapsenberg, S. Ray Smith, Jerome Magnusen, Steve Reid, Robin Lamp, Devesh Singh, Jerry Hall, Trent Tate, Krista Lea, James Dedecker, and Gabriel Roberts[1]

Abstract

Heading date is a critical management point in production of grass hay, haylage, and pasture because of its relationship to forage quality, but the forage grass industry does not have a standardized maturity rating system to allow comparison of heading dates across species and varieties. The AFGC Forage Grass Maturity Working Group was formed in 2016 to develop a grass maturity index to fill this need. Trials were run between 2017 and 2023 to evaluate relative heading dates among 70 cool-season grass varieties in six locations across the primary US cool-season grass growing region. Relative heading dates were consistent across site-years, allowing identification of nine reference varieties for a maturity index.

Introduction: Heading date is a critical maturity factor for selection in the development of new forage grass varieties because stage of maturity is the single most important factor influencing the nutritional quality of forages. As forage matures, yield typically increases but cell walls thicken and lignify, increasing indigestible fiber content while protein and digestible nutrient content declines. The rate of maturation in grasses is driven by both genetics and environmental factors such as photoperiod, temperature, and length of growing season (Robins et al., 2012; 2017) and understanding of grass heading date is essential to effective management of forage yield and quality, especially in mixtures where compatibility of maturity dates is essential to maintenance of harvested quality.

There is no uniform industry rating system in place for describing and comparing the heading date of forage grasses. This contrasts with well-defined and objective maturity rating systems for other major crops such as alfalfa (fall dormancy), corn (relative maturity rating), and soybeans (maturity group). In forage grasses, each seed distributor makes their own independent determination of heading dates for their varieties. This leads to inconsistent information across the marketplace, making it difficult for hay growers and livestock producers to select grass species or varieties with predictable maturity dates relevant to their harvest goals for hay, haylage, or pasture. The high risk level associated with making a poorly informed variety choice makes producers reluctant to invest in improving their grasslands through better plant genetics. Lack of an industry standard for maturity also makes it difficult for grass breeders to make meaningful progress in developing varieties with reliable heading dates relative to others on the market.

The AFGC Forage Grass Maturity Working Group was formed in 2016 in response to this challenge and tasked with developing a forage grass maturity index. The first step of this

[1]Michigan State University, cassida@msu.edu; Jethro Ag Services; University of Kentucky; GO Seed; Barenbrug USA; DLF USA; GO Seed; Wilbur Ellis; DLF USA, University of Kentucky; Michigan State University; and University of Kentucky; respectively.

challenge was determining whether cool-season grass heading dates rank consistently across diverse environments. Previous research in this area is focused primarily on orchardgrass, which does not always exhibit consistent maturity rankings across environments (Robins et al., 2012). Our approach differs because we hypothesized that including multiple species allows comparison across a wider range of variety maturities than is possible within a single species and will produce a more robust comparison of relative maturity.

Our objectives were: 1) determine whether cool-season grass variety heading dates rank in consistent order across diverse locations and multiple years, and 2) identify a subset of consistently ranked varieties to be used as reference standards for a maturity ranking index.

Materials and Methods: Test plots were grown from 2017 through 2023 in locations selected to provide a range of environments across the typical cool-season grass usage range in the US. Site characteristics are summarized in Table 1. The trial protocol followed the general principles used for GVRB/PVP applications (AOSCA, 2020). Seventy grass varieties were selected to represent a range of potential heading dates across eight common species and included: tall fescue (*Schedonorus arundinacea,* 15 varieties), orchardgrass (*Dactylis glomerata*, 16 varieties), timothy (*Phleum pratense*, 10 varieties), bromegrass (*Bromus* spp, 5 varieties), perennial ryegrass (*Lolium perenne*, 11 varieties), annual ryegrass (*L. multiflorum*, 6 varieties), hybrid ryegrass (*Lolium* spp., 5 varieties), and Kentucky bluegrass (*Poa pratensis*, 2 varieties). All seed was held at Univ. Kentucky for distribution to test sites after blind coding so that evaluators could not be biased by expectations of performance from any variety. Weather data were obtained from regional weather stations as close as possible to each trial site.

Single seedlings were started in plug trays in greenhouses by each cooperator and transplanted into spaced-plant nurseries as soon as vigorous root systems developed in the plugs (about 45 days after seeding). Each spaced-plant nursery was designed as a randomized complete block with three replications. Each experimental unit contained 10-25 individual plants of a given variety spaced 2 feet apart, with 3 feet between rows. Nurseries were fertilized to provide N, P, and K according to local soil tests and fertility recommendations. Weeds were controlled using a combination of herbicides, cultivation, and hand-weeding as needed. No maturity data was collected during the establishment year.

Heading dates were recorded in the first cycle of growth in the first and second years after establishment. Plots were observed three times per week once heading began. Heading date for each individual plant was recorded as the day of the year when at least three heads were completely emerged from the flag leaf collar, which corresponds to the R1 phenological stage described by Moore et al. (1991). Heading dates were averaged across all plants in each experimental unit and means were used in statistical analyses. After heading data was collected each spring, plots were periodically mowed to a 4-inch height to control vegetation during the season.

Statistical analysis was conducted using SAS (SAS, Inc., Cary, NC) to determine the consistency of rank order of variety heading dates within and across locations. Data were analyzed using Spearman's Rank Correlation to determine whether variety heading dates were consistently ranked among site-years.

Reference varieties were selected using an iterative process based on identifying varieties with consistent relative rankings for heading date across site-years. Prospective varieties for a 10-point rating index were chosen from the ranked order of mean heading dates across all site-years using a 5- to 7-day interval across the observed range. Varieties were dropped from consideration if: 1) they had too much missing data due to lack of inclusion in plots or poor persistence; 2) did not maintain consistent relative ranking across all site-years; or 3) were at or near the end of their commercial lifespan. These steps were repeated until a final set of reference varieties were agreed on by the research team.

Results and Discussion: Spearman's rank correlation showed that variety heading dates consistently ranked similarly ($P < 0.05$) within all possible pairs of site-years (Fig. 1), with two exceptions. First, heading dates in Oregon in 2021, where severe drought occurred during spring growth and heading measurements, did not align with most other site-years. Second, comparisons with less than 21 varieties in common across a pair of site-years generally failed to align.

The ranked order of mean heading dates across all varieties and site-years is shown in Fig. 2. Across all site-years, mean heading dates exhibited a 58-day range from earliest to latest variety. There was considerable overlap in heading dates among species and this supported our approach of selecting reference varieties across species instead of within species.

Of 16 total site-years, 10 met all iterative criteria and were used for reference variety selection. Data from one site-year was not collected due to the covid interruption. Reasons for excluding other site-years or varieties from consideration included: site-years that did not establish all species groups; failure of individual varieties to persist for the full two years and provide at least 10 total observations per site-year; and atypical growing seasons. A few varieties did not maintain usable seed germination over the entire time frame of the tests and were not included in later tests. The iterative selection method for reference varieties resulted in identification of nine varieties with heading dates that ranked similarly across each site-year (Fig. 3). Reference varieties in the center of the ranges were markedly consistent in relative rank across environments, while it was more difficult to maintain relative rank at the early and late edges of the range.

In summary, heading dates of cool-season grass varieties maintained generally consistent rankings from early to late across the six locations studied. The working group selected nine varieties as potential reference varieties for a forage grass maturity index rating. A proposal for a testing protocol is presented in a companion paper in these Proceedings (Smith et al., 2024).

Literature Cited:

Assoc. of Official Seed Certifying Agencies. 2021. Online https://www.aosca.org/grass/. Verified 11/30/23.

Moore, K.J., L.E. Moser, K.P. Vogel, S.S. Waller, B.E. Johnson, and J.F. Pedersen. 1991. Describing and quantifying growth stages of perennial forage grasses. *Agronomy J.* 83:1073-1077.

Robins, J.G., S.R. Smith, M.H. Hall, C.D. Teutsch, and D.J. Undersander. 2017. Associations among U.S. locations for orchardgrass production. *Crop Forage & Turfgrass Management.* Vol 3, online Dec 2017.

Smith, S.R., K. Cassida, R. Kapsenberg, J. Hall, R. Lamp, J. Magnuson, T. Tate, D. Singh, S. Reid, K. Lea, J. Dedecker, G. Roberts. 2024. Protocol for evaluation of forage grass maturity. Proc. American Forage & Grassland Conference, Mobile, AL. Jan. 7-10, 2024.

Table 1. Characteristics of sites used for selection of reference varieties for a cool-season grass maturity index.

	Participant					
	BAR[a]	DLF	DLF	GO	MSU	MSU
State	OR	OR	KY	IA	MI	MI
City	Albany	Philomath	Berry	Richland	East Lansing	Chatham
# of site-years	6	2	1[b]	2	2	2
Latitude	44°, 37.548'N	44°, 31.769'N	38°, 32.598'N	41°, 11.093'N	42°, 42.471'N	46°, 20.692'N
Longitude	122°, 55.985'W	123°, 21.406'W	84°, 23.896'W	91°, 59.512'W	84°, 28.090'W	86°, 54.975'W
Elevation (ft)	272	294	645	771	858	878
Precipitation (inch/yr)	44	59	43	37	34	33
Frost-free days	189	189	179	174	153	123

[a]BAR-Barenbrug USA; DLF – DLF Pickseed USA; GO – Go Seed; MSU – Michigan State Univ.
[b]One site-year lost during covid 19.

	BAR1-OR-18B	BAR1-OR-19A	BAR2-OR-20B	BAR2-OR-21A	BAR3-OR-22B	DLF-KY-17A	DLF-KY-18B	DLF-OR-19A	GO-IA-17A	GO-IA-18B	MSU-EL-21A	MSU-EL-22B	MSU-CH-22A	MSU-CH-23B
BAR1-OR-17A	0.67*** 26	0.74*** 24	0.70*** 24	0.51 14	0.16 14	0.69*** 26	0.28 16	0.38 20	0.46* 25	0.78*** 24	0.64*** 25	0.68*** 25	0.59** 22	0.59** 22
BAR1-OR-18B		0.94*** 24	0.99*** 24	0.46 14	0.89*** 14	0.95*** 26	0.94*** 16	0.97*** 20	0.83*** 25	0.82*** 24	0.96*** 25	0.99*** 25	0.98*** 22	0.96*** 22
BAR2-OR-19A			0.87*** 64	0.64*** 53	0.77*** 54	0.8*** 64	0.62*** 35	0.78*** 31	0.64*** 59	0.72*** 51	0.82*** 64	0.86*** 55	0.75*** 56	0.78*** 47
BAR2-OR-20B				0.34* 53	0.97 54***	0.88*** 65	0.88*** 35	0.75*** 31	0.79*** 59	0.83*** 51	0.85*** 64	0.92*** 55	0.81*** 56	0.79*** 47
BAR3-OR-21A					0.31* 53	0.52*** 53	0.47** 34	0.30 26	0.12 48	0.31* 42	0.29* 53	0.40** 45	0.20 47	0.28 39
BAR3-OR-22B						0.88*** 54	0.83*** 34	0.45* 26	0.74*** 49	0.8*** 42	0.82*** 54	0.85*** 45	0.77*** 48	0.64*** 39
DLF-KY-17A							0.71*** 38	0.45** 32	0.72*** 62	0.84*** 0.84	0.74*** 65	0.81*** 56	0.66*** 57	0.63*** 48
DLF-KY-18B								0.70*** 27	0.41* 34	0.42 33	0.67*** 35	0.81*** 35	0.46** 32	0.51** 32
DLF-OR-19A									0.51** 30	0.43* 29	0.90*** 32	0.91*** 32	0.86*** 29	0.88*** 29
GO-IA-17A										0.76*** 54	0.74*** 60	0.76*** 53	0.75*** 54	0.69*** 45
GO-IA-18B											0.79*** 52	0.84*** 52	0.77*** 47	0.76*** 44
MSU-EL-21A												0.96*** 56	0.94*** 57	0.97*** 48
MSU-EL-22B													0.93*** 51	0.95*** 48
MSU-CH-22A														0.97*** 48

Key for each box
Spearman's correlation coefficient
Number of varieties compared

*,**, ***Significant at $P < 0.05$, 0.01, and 0.001, respectively.

Fig. 1. Spearman's Rank Correlation and number of varieties compared for mean heading date of cool-season grass varieties evaluated across 15 site-years in Oregon, Iowa, Kentucky, and

Michigan between 2017 and 2023. Site-year labels indicate "test number-site-year," with A indicating first year of heading data and B second year for that test.

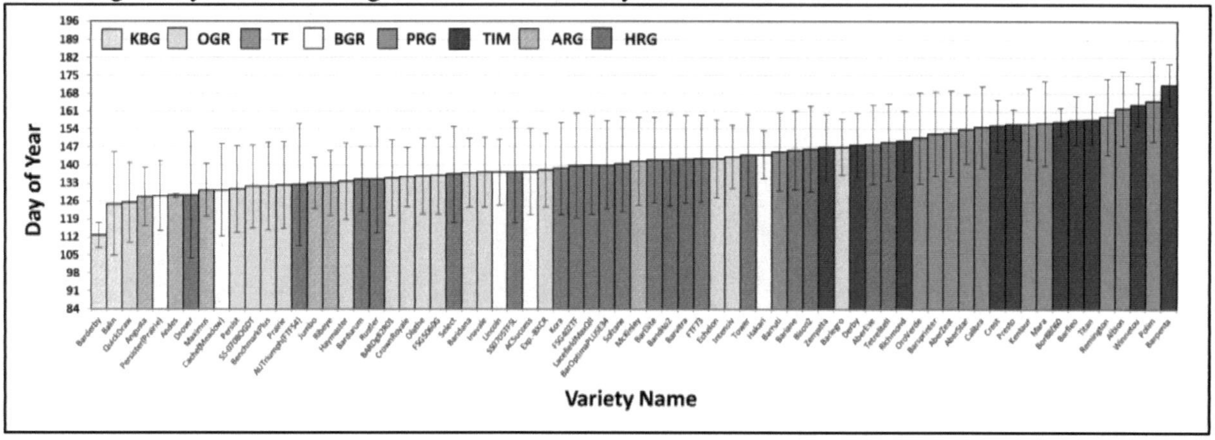

Fig. 2. Mean heading date and standard deviation for 70 cool-season grass varieties in eight species groups (KBG - Kentucky bluegrass, OGR - orchardgrass, TF - tall fescue, BGR - *Bromus* spp., PRG – perennial ryegrass, TIM – timothy, ARG – annual ryegrass, HRG – hybrid ryegrass) evaluated across 15 site-years in Oregon, Iowa, Kentucky, and Michigan between 2017 and 2023.

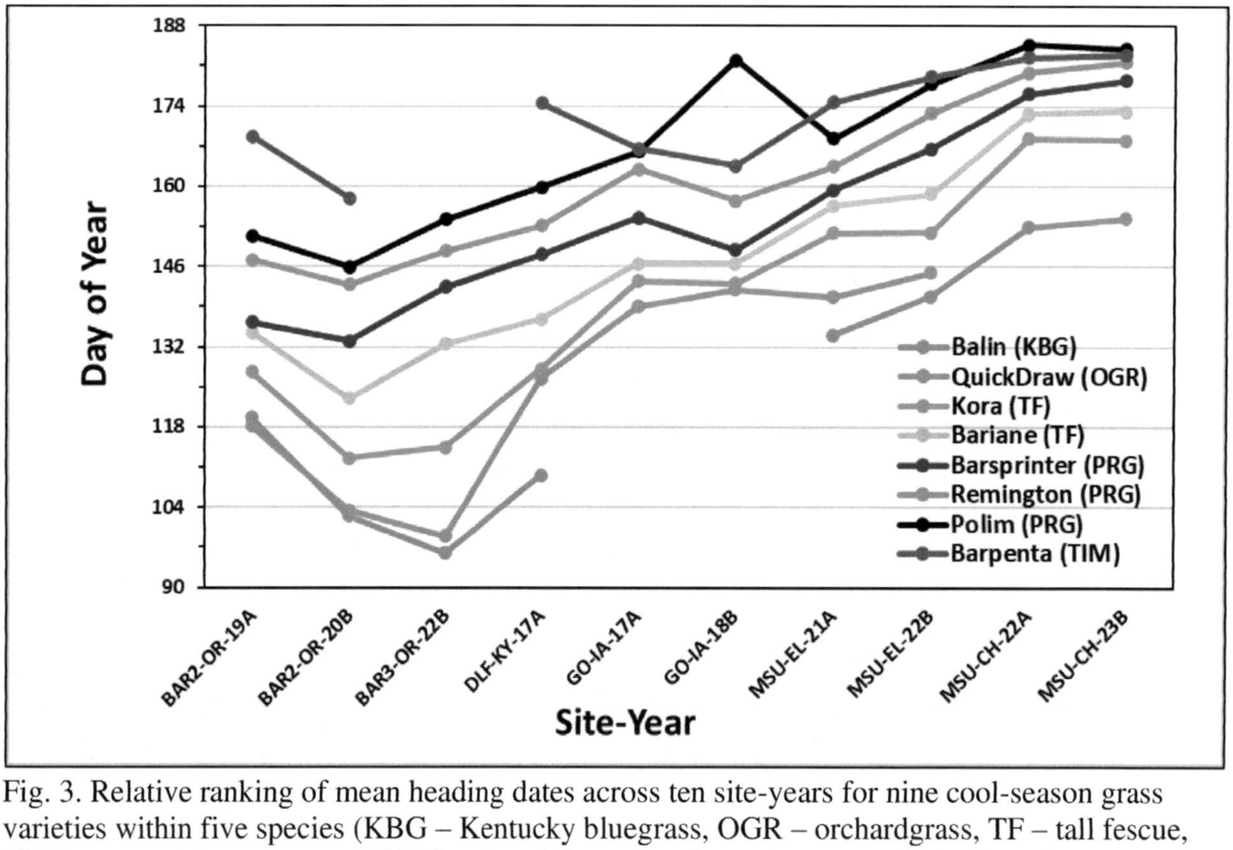

Fig. 3. Relative ranking of mean heading dates across ten site-years for nine cool-season grass varieties within five species (KBG – Kentucky bluegrass, OGR – orchardgrass, TF – tall fescue, PRG – perennial ryegrass, and TIM – timothy) proposed as reference varieties for a forage industry standard cool-season grass maturity index. Site-year labels indicate "test number-site-year," with A indicating first year of heading data and B second year for that test.

IMPACT OF DIFFERENT SPINY PIGWEED INFESTATION LEVELS IN GRAZING SYSTEMS

L.M.D. Queiroz, J.C.B. Dubeux Jr., K.T. Oduor, I.L. Bretas, K.R. Trumpp, J.P. Acuna, M.A. Bernardini, B.E. Bizzuti-Cremostim, F.F. Simili, M.A. Lira Jr., and M. Ruiz-Moreno[1]

Weeds can be a major problem for livestock farms as they compete with grass for vital resources. This competition can lead to a decline in forage productivity, resulting in economic losses. To better understand the impact of spiny pigweed (*Amaranthus spinosus*) on common bermudagrass (*Cynodon dactylon*) pastures, this study evaluated the effect of three different levels of infestation on both herbage and animal responses. The study was conducted at NFREC in Marianna, FL. Duracor herbicide from Corteva was used to impose three infestation levels. Herbage mass and accumulation were measured every 21-d using the double sampling method and exclusion cages. Additionally, hand-plucked samples were collected for nutritive value analysis. Animals were weighed every 21-d and put-and-take animals were used to maintain similar herbage allowance among treatments. The presence of spiny pigweed in the pasture affected the digestibility of bermudagrass, reducing it from 47 to 42%. The crude protein concentration of bermudagrass varied throughout the season, with lower values observed during the last two evaluations. However, pigweed encroachment did not have any significant impact on the average daily gain, gain per area, and stocking rate, which averaged at 0.3 kg/hd, 133 kg/ha, and 2.1 AU/ha, respectively.

[1]Ph.D. Student (ldantasqueiroz@ufl.edu); Professor, Forage-Agronomy (dubeux@ufl.edu) Ph.D. Student (kenneth.oduor@ufl.edu); Post-doctoral associate (ig.limabretas@ufl.edu); MS Student (ktrumpp@ufl.edu); MS Student (portuguez.javier@ufl.edu); MS Student (m.araujobernardi@ufl.edu); Exchange Ph.D. Student, (beatrize.bizzuti@ufl.edu); Visiting Professor (similif@ufl.edu); Visiting Professor (lirajr.ma@ufl.edu); Biological Scientist (ruizm@ufl.edu) – Noth Florida Research and Education Center, Marianna, FL.

FORAGE ACCUMULATION AND NUTRITIVE VALUE CHARACTERISTICS OF STOCKPILED BAHIAGRASS IN ALABAMA

J.B. Elmore, M. Agar, P. Zessin, K. Palmer, A. Wiggins, K. Kelley,
S.L. Dillard and M.K. Mullenix[1]

Bahiagrass is a perennial warm-season grass widely used in Alabama for hay production or grazing in the central and southern portions of the state. Identifying alternatives to extend the production season of bahiagrass pastures can provide supplemental forage and fill a production gap in the fall. Stockpiling extends forages for later use, but most research has focused on tall fescue, bermudagrass, and more recently, limpograss. We evaluated bahiagrass under three stockpiling strategies (6-, 8-, and 10-weeks of accumulation prior to the first anticipated frost date) and three nitrogen fertilization strategies [no N, split application of N (30 lb/A at initiation of stockpiling and 30 lb/A 28-d from first application), or 60 lb/A] for fall, 2021 and 2022) at four locations. Stockpiling was initiated on Sept 15 in each year of the study. Data collected during the stockpiling period included: forage mass and forage nutritive value (digestibility, crude protein, and fiber fractions). Forage accumulation ranged from 1,110 to 1,548 lb DM/A across stockpiling period length and N fertilization strategies. Stockpiling was feasible for 6 to 8 weeks after initiation in Clanton, AL but frost significantly reduced available forage at 10-week. N application increased forage accumulation by 250 to 300 lb DM/A across all periods. Forage CP ranged from 12 to 14% across N strategies, while TDN averaged 61% at the 6- and 8-week periods, and 59% at the 10 week interval. Forage NDF and ADF did not differ at 6 or 8 weeks, but increased at 10 weeks. Stockpiled bahiagrass supported moderate forage production during the fall with nutritive value sufficient to meet most nutritional needs in beef cow/calf operations.

[1]Joshua B. Elmore, Regional Extension Agent, Alabama Cooperative Extension System, Clanton, AL (elmorjb@auburn.edu).

ENHANCING FORAGE SYSTEM PROFITABILITY AND SUSTAINABILITY WITH POLYSULPHATE FERTILIZER

AJ Foster[1], Jason Haegele[2], Christi Falen[3], Mark Wiesepape[4], and Hal Russell[4]

Polyhalite is a new commercially available fertilizer offered as Polysulphate by ICL. This mineral fertilizer is a sustained release source of potassium (14% K_2O), calcium (12.2 % Ca), magnesium (3.6% Mg), and sulfur (19.2% S). In previous research trials, Polysulphate has increased both yield and quality in forage systems. For example, in Missouri, USA, Polysulphate as a sulfur source increased protein by 3% compared to ammonium sulfate, and in the Netherland and Finland, Polysulphate increased dry matter yields by 10 and 3%, respectively, over the grower standard practice. In an alfalfa trial from Wisconsin, USA, application of Polysulphate increased dry matter yield, protein content, and digestibility compared to the grower standard blend of ammonium sulfate and potash. Polysulphate offers the opportunity of a sustained-release fertilizer that can provide adequate supply of four nutrients in a single granule that are important for profitable yield and quality in forage systems.

[1]ICL Growing Solutions Agronomy Technical Services Manager – Southern US, [2]ICL Growing Solutions Agronomy Lead – North America, [3]ICL Growing Solutions Agronomy Technical Services Manager – Western US, [4]ICL Growing Solutions Technical Sales Manager.

SOIL-TEST BIOLOGICAL ACTIVITY UNDER FORAGE AND GRAZING LANDS ON PRIVATE FARMS IN NORTH CAROLINA AND VIRGINIA

A.J. Franzluebbers[1]

Forage and grazing lands can be managed to restore soil health conditions, but quantitative information across a diversity of farms and physiographic regions is lacking. During the winters of 2021/22 and 2022/23, soil was sampled at 0-10-cm depth across Virginia and North Carolina, respectively, to assess soil-test biological activity as an indicator of soil life. Grazed pastures, hayfields, woodlands, and croplands were sampled on 238 farms. Soil texture, physiographic region, and management history were important variables that affected soil-test biological activity, i.e. the flush of CO_2 during 3 days following rewetting of dried soil. Soil-test biological activity was 195 $\pm$ 11 mg CO_2-C kg^{-1} 3 d^{-1} (mean $\pm$ standard error) in the Coastal Plain of North Carolina (n=145), 390 $\pm$ 7 mg CO_2-C kg^{-1} 3 d^{-1} in the Piedmont of North Carolina (n=393), 495 $\pm$ 12 mg CO_2-C kg^{-1} 3 d^{-1} in the Blue Ridge of North Carolina (n=145), and 439 $\pm$ 10 mg CO_2-C kg^{-1} 3 d^{-1} under grasslands in the Piedmont, Great Valley, and Blue Ridge regions of Virginia (n=186). Values >300 mg CO_2-C kg^{-1} 3 d^{-1} are considered adequate, and most samples exceeded this level, suggesting high soil health condition. These results can be interpreted to assess soil nitrogen supplying capacity, as well as overall soil health condition.

[1]Reseach Ecologist, USDA-Agricultural Research Service, NCSU Campus Box 7620, Raleigh NC 27695 (alan.franzluebbers@usda.gov).

EXPANDING OPTIONS FOR ESTABLISHMENT METHODS OF NATIVE WARM-SEASON GRASSES IN EASTERN OHIO

C.H. Gelley, C.Y. Rodríguez Hernández, and M.B. Chiavegato[1]

The establishment of Native Warm-Season Grasses for conservation and forage purposes is appealing to many Ohioans but is often challenging on farm due to weed pressure. Current recommendations for successful establishment include the strategic use of herbicides to combat weeds, but many producers would prefer to employ methods with little to no herbicide applied. In our investigation from 2021 to 2023, three weed control treatments for were utilized and compared for newly seeded stands of big bluestem and indiangrass, switchgrass, and eastern gamagrass in a randomized block design with four replications at the Eastern Agricultural Research Station in Belle Valley, Ohio. The first treatment began with a burndown herbicide application before planting and follow-up applications of broadleaf herbicide as needed. The second treatment began with a burndown herbicide application before planting a winter cover crop, followed by chemical termination, then planting of the forages, with follow-up applications of broadleaf herbicide as needed. The third treatment utilized 6" rotary tillage, a summer cover crop, a winter cover crop, then planting of the forages, with no herbicide applied in the process. Weed versus forage frequency metering, forage yield, and forage quality were monitored from 2021-2023 to determine which treatment most effectively aided forage establishment. Results indicated that treatment three was not effective for weed suppression in this case, while treatments one and two led to increased frequency of native forages versus weeds stands over time.

[1]Extension Educator, The Ohio State University Department of Extension, Caldwell, OH (gelley.2@osu.edu); M.S. Candidate, The Ohio State University Department of HCS, Columbus,

EFFECT OF AN IMPROVED GRAZING MANAGEMENT SYSTEM ON DAIRY HEIFER PERFORMANCE

A.M. Grev, S.B. Potts, J.W. Semler[1]

The objectives of this study are to investigate the effects of improved grazing management on pregnant dairy heifer performance and its economic feasibility relative to a conventional management system. Heifers at the Central Maryland Research & Education Center were utilized on a rolling basis, with heifers added following pregnancy confirmation and removed three weeks prior to calving. Heifers were divided into two treatment groups which varied in size throughout the study period (15 to 22 per group) but were kept consistent between treatments. Heifers in the CON group received a daily TMR and had access to one continuously managed 5-acre perennial pasture. Heifers in the ROT group were rotationally grazed across 19 acres of both perennial and annual pastures subdivided into approximately 0.5-acre paddocks (1-2 d rotation); ROT heifers also received a daily corn/mineral mix (1 lb/head/d) to ensure dietary requirements were met. Both groups of heifers were measured [BW, BCS, hip height (HH)] and fecal samples were collected on a bi-weekly basis throughout the grazing season. Initial BW, BCS, and HH did not differ between treatment groups. Final BW and HH did not differ, but final BCS was greater for CON heifers compared to ROT heifers. For both groups, heifer average daily gain (ADG) varied considerably across weigh periods, but ADG was greater for CON compared to ROT. Results indicate a slight growth disadvantage for heifers in the ROT system; however, heifer performance was still good and expenses for this system may be reduced compared to a conventionally managed system.

[1]University of Maryland Extension, WMREC, Keedysville, MD (agrev@umd.edu).

EFFECT OF NITROGEN SOURCES AND RATES ON DRY MATTER YIELD OF HYBRID BERMUDAGRASS

Danny Griffin, John Jennings, Leo Espinoza, Kenny Simon, Andrew Bolton, Matthew Davis, Brett Gordon, Brian Haller, Amy Heck, Michelle Mobley, Kyle Sanders, and Andrew Sayger

The effectiveness of different nitrogen fertilizer sources for summer forage production is often discussed. There is debate whether urea nitrogen is less effective than ammonium nitrate for bermudagrass production. Most research shows either no or small differences in forage yield between these sources, yet the belief persists that much of the N in urea will be lost during hot summer weather. Further, new additives are available that reduce N volatilization losses from urea potentially improving its effectiveness for summer forage production. This study was initiated to compare the forage yield response of hybrid bermudagrass to ammonium nitrate, urea, urea with two N loss preventative additives, and urea ammonium nitrate (UAN 32%) applied at three N rates. Treatments were arranged as a randomized complete block design with four replications. The plots were harvested 4 times and clipped to a height of approximately 3 inches. Nitrogen sources were not significantly different for dry matter yield. Nitrogen rate did significantly influence yield with dry matter increasing with increasing N rate. The biggest DM increase per unit of N was from the 30 lb N per acre rate followed by the 60 lb N rate. It should be noted that each nitrogen application received rainfall within 5 days after nitrogen application.

YIELD POTENTIAL OF LATE SUMMER PLANTED OATS FOR FALL GRAZING

C. Grimes[1], K. Simon[2], B. Finch[3], and D. Kennedy[4]

Producers are looking for opportunities to reduce the amount of hay required for their operations. One possibility that has sparked interest among growers was to extend the grazing season further into the fall and possibly early winter by planting oats in late summer. Previous studies have found that late summer planted Jerry oats have consistently provided more forage yield in the fall than winter wheat or cereal rye. However, limited information is available comparing late summer forage yield of oat varieties. This study was established to determine the yield potential of late summer planted oats for fall grazing. The study consisted of forage and black oat varieties that were selected based on local availability. The forage varieties planted were Goliath, Jerry, Magnum, and Pearl. The black oats varieties planted were Cosaque and Soil Saver. Plots were no-till planted into sod-suppressed bermudagrass on September 1, 2022, at a 100lb/A seeding rate. Fertilizer was applied at planting according to soil test recommendations. Canopy closure, canopy height, and forage yield were measured on December 1, 2022. Magnum, Pearl, and Goliath had greater canopy closure than the other varieties, while Jerry oat had the most upright growth with a 16" average canopy height. Forage oat varieties produced more dry matter (DM) biomass than the black oat varieties, yielding an average of 1,710 lb. DM/A compared to 1,135 lb. DM/A, respectively.

[1]Craighead County Agent, University of Arkansas System Division of Agriculture, Cooperative Extension Service, Jonesboro, AR (cgrimes@uada.edu); [2]Forages Instructor, University of Arkansas System Division of Agriculture, Cooperative Extension Service, Little Rock, AR (ksimon@uada.edu) [3]Soil Fertility Specialist, University of Arkansas System Division of Agriculture, Cooperative Extension Service, Little Rock, AR (bfinch@uada.edu); [4]Associate Dean and Professor of Animal Science, Arkansas State University, Jonesboro, AR (dkennedy@astate.edu).

EFFECTS OF EXERCISE WHEN TRANSITIONING FROM FREE-CHOICE TALL FESCUE (*LOLIUM ARUNDINACEUM*) GRASS DIET IN LATE SUMMER TO FREE-CHOICE TALL FESCUE HAY DIET IN LATE FALL ON BLOOD METABOLITE CONCENTRATIONS, BODY WEIGHT, AND BODY CONDITION SCORES OF STOCK HORSE MARES AND GELDINGS

Phillip Gunter, Riley Brown, and Stephen King[1]

Equine exercise may influence horse metabolic and physiological functions. Limited research has been conducted on the impact of diet transition from actively growing forage to hay diets and exercise regimen on equine metabolic function. The current study utilized eight horses that were split into two groups (n = 4/group); Worked horses were exercised three times a week, and Unworked horses were not exercised. All horses had ad libitum access to tall fescue pasture (d 1 – 71) or tall fescue hay (d 72 – 127) for a total of 127 grazing-days. Pasture samples were collected monthly and hay samples were collected by lot for forage nutritive value analysis. Blood samples were collected every two weeks to determine plasma-fructosamine concentrations and insulin levels, and horses were weighed on a standing scale and evaluated for body condition score (BCS) at the same time as blood draws. Date were analyzed using PROC MIXED of SAS (ver. 9.4) with $\alpha \leq 0.05$. Insulin concentration ($P < 0.05$) and BCS ($P < 0.05$) were affected by exercise regimen, while plasma fructosamine ($P > 0.05$) and body weight ($P > 0.05$) were not affected by exercise. Results are interpreted to mean that horses under an exercise regimen during the diet transition from pasture to stored forage had lower insulin concentrations and lower BCS than unworked horses. While body weight did not differ, the change in BCS may indicate a change in body composition between the groups. Further research with more animals would be required to fully elucidate these differences.

[1]Dept. of Agriculture and Food Science, Western Kentucky University, Bowling Green, KY.
Corresponding author: Phillip Gunter – phillip.gunter@wku.edu.

STOCKPILING WARM SEASON GRASSES TO REDUCE WINTER FEED COSTS

B.W. Haller[1] and K.J. Simon[2]

The cost of hay and feed for winter feeding is the largest expense of maintaining a beef cattle herd. Estimates in Arkansas in 2021 - 2022 according to USDA-NASS showed that hay prices averaged $133/ton. All too often, producers finish harvesting hay in the fall and then begin feeding it soon afterward. Adopting pasture management practices that extend the grazing season avoids investing more cost into forage that could be grazed instead of being harvested for hay. Stockpiling warm season grass is a pasture management practice that has the potential to reduce winter feed cost for beef producers. A stockpiled warm season grass demonstration was conducted on forty acres in the Fall/Winter 2022 through Winter 2023. The fields were clipped to 2-3 inches in height in early August and fertilized with 125 lb/A of urea fertilizer on August 4th. Grazing was deferred until November 13, 2022. The stockpiled forage was strip grazed. Ninety-eight animal units (AU = 1,000 lb cow) grazed on the stockpile for 48 days. The forage quality of the stockpiled was 7.6% CP and 55.5% TDN. Protein tubs were used as a supplement to balance animal nutritional requirements. The total savings between strip grazing the stockpiled forage vs. feeding hay was $6,854.66 or $98.91 per animal unit.. Stockpiling warm season grass will save producers money.

[1]County Extension Agent – Staff Chair, University of Arkansas Division of Agriculture Cooperative Extension Service White County, Searcy, AR, (bhaller@uada.edu); [2]Instructor-Forages, University of Arkansas Division of Agriculture Cooperative Extension Service, Little Rock, AR, (ksimon@uada.edu).

GRAZING MANAGEMENT STRATEGIES TO ADD VALUE TO MARKET COWS

T.A. Hamilton[1], G.E. Aiken[1], M.D. Flythe[1], D. Miller[2], E.S. Flynn[3], J.R. Weinert-Nelson[1], and B.E. Davis[1]

Approximately 40% of the U.S. cow herd is located in the "fescue belt", with 15 – 20% turnover expected annually. Adding condition to market cows can improve economic returns by increasing cutability and total carcass weight. Chemical seedhead suppression with Chaparral™ herbicide (CH) and overseeded red clover (RC) in endophyte-infected tall fescue (E+ TF) grazing systems increases steer average daily gain (ADG). The objective of this study was to evaluate pasture management strategies to improve performance and profitability of market cows grazing E+ TF. Market cows (n = 24) were purchased in mid-May 2018 and 2019. Cows were blocked by body weight (BW) and body condition score (BCS) into 6 grazing groups which were assigned to a rotational block of 2.0-acre paddocks with 3 treatments: 1) E+ TF control, 2) E+ TF with CH (applied at 2.0 oz/A; 1-May), or 3) RC (frost-seeded, 5.0 lb/A, 25% establishment). The cows grazed treatments for 67-69 days, with BW and BCS collected at the start and termination of grazing. Cows were harvested through Beef Solutions, LLC (Kentucky Beef Network, Lexington, KY). Cows grazing both CH (1.95 lb/d; +1.0 BCS) and RC (1.81 lb/d, +1.1 BCS) had increased ADG and BCS in comparison to E+ TF control (1.36 lb/d, +0.88 BCS. Similarly, CH (+$203.00) and RC (+$225.30) increased total net profit compared to cows grazing E+ TF control pastures ($152.57). Grazing for a short period could provide an opportunity for producers to add value to market cows. Implementing strategies like CH application and overseeding RC can further improve performance and value of market cows grazing E+ TF pastures.

[1]USDA-ARS Forage Animal Production Research Unit, Lexington, KY; [2]Kentucky Beef Network, Lexington, KY; [3]Corteva Agriscience, Indianapolis, IN.

FORAGE MANAGEMENT EDUCATION THROUGH FEATURE ARTICLES

G.J. Hannan, K.D. Johnson, and K.L. Smith[1]

The Indiana Forage Council has a quarterly newsletter that highlights their members through feature articles. The leadership team developed an opportunity for a Purdue University student to receive academic credit for article preparation. Before interviewing the featured members, Kevin Leigh Smith, an advisor and communication specialist from the Agricultural Sciences Education and Communication Department, gave the student a brief course on feature article writing. In the summer of 2022, two forage-livestock operations were visited. Ronnie Boehm custom grazes beef cattle on his family farm in Spencer County, IN. The other interviewee was Jason Tower, Superintendent of the Southern Indiana Purdue Agricultural Center in Dubois County, IN. During the visits, photos, audio and video recordings, and handwritten notes were taken. The student prepared the feature articles, and Smith reviewed them for technical content and writing effectiveness. Visiting the forage council members at their operations provided an excellent and fun learning experience. This experience reinforced many of the principles learned in the Purdue University forage management class and deepened knowledge of pasture management practices. State forage councils may improve their ability to promote forages and help producers learn from the examples of others by having a university student prepare a feature article for newsletters or other communication methods.

[1]Masters Graduate Student, Purdue University Agronomy Department, West Lafayette, IN (ggresham@purdue.edu); Professor of Agronomy, Purdue University, West Lafayette, IN (johnsonk@purdue.edu); Continuing Lecturer/Advisor/Communication Specialist, Purdue University Department of Agricultural Sciences Education and Communication, West Lafayette, IN (smith79@purdue.edu).

FALL OR SPRING NITROGEN AND SULFURS EFFECT ON CEREAL RYE

J.M. Hartschuh and A.M. Gahler[1]

Nitrogen and sulfur are critical elements that drive cereal grain forage yield and crude protein. This study was conducted to determine the effects of fall-applied nitrogen and spring-applied nitrogen and sulfur on winter annual cereal rye forage yield and nutritional value. Small plots in a randomized complete block split block design were planted in the fall of 2021 and 2022 to variety not stated cereal rye. Fall treatments were applied at planting of either no fall nitrogen or 20 lb/A of fall nitrogen as urea. Spring treatments were applied at green up in early April of 50 or 70 lb/A of nitrogen as urea and 0 or 20 lb/A of sulfur as ammonium sulfate with nitrogen value accounting in reduced urea applications. 2022 harvested plots were hand harvested while 2023 plots were harvested with a small plot forage harvester harvesting a 3-foot pass with both years being harvested in the boot stage. Samples were weighed and then dried for laboratory quality analysis. 2023 forage dry matter yield was significantly higher than 2022 while crude protein was lower in 2023 making year a significant factor, but there was an interaction between nitrogen and sulfur treatment and year. Regardless of the spring nitrogen rate, the fall nitrogen application of 20 lb/A out yielded the no fall nitrogen treatments by over one and a half tons in 2023 and over a quarter ton in 2022. Spring nitrogen treatments did not affect yield in 2022 while in 2023, 70 lb/A of nitrogen yielded a half-ton more than the 50 lb/A nitrogen treatment. Sulfur treatments did not significantly increase yield or quality.

[1]Field Specialist Dairy Management and Precision Livestock, The Ohio State University Extension, Fremont, OH (Hartschuh.11@osu.edu); Extension Educator, The Ohio State University Extension, Fremont, OH (Gahler.2@osu.edu).

NITROGEN RATE AND TIMING EFFECTS ON THE YIELD AND NUTRITIVE VALUE OF TALL FESCUE

B.L. Hendrix, C.D. Teutsch, and J.H. Grove[1]

Late season nitrogen (N) fertilization is a common practice in turfgrass production, resulting in an extended green period in the fall without excessive shoot growth. This longer green period results in higher levels of carbohydrate reserves and enhanced root growth compared with spring-summer N fertilization. However, limited research has considered the impact of fall N application on grass growth in hayfields and pastures. Our objective was to determine the impact of fall and spring N applications on the dry matter yield and nutritive value of tall fescue. This experiment was conducted at the University of Kentucky Research and Education Center (UKREC) in Princeton, KY in 2022-23. Plots received N either in the fall, in the spring, or in both fall and spring. Fall rates were 0, 40 or 80 lb N/A, and spring rates were 0, 40, 80 or 120 lb N/A, all as urea (46-0-0). Plots were harvested in early May and again in mid-July using a forage harvester equipped with load cells. Subsamples were collected for dry matter and nutritive value analysis at both harvests. The data indicate that fall N increased first harvest yield but had a limited impact on second harvest yield. Total yield exhibited a linear response up to 200 lb total N/A. The spring N rate impacted CP, ADF, and NDF values, but these same constituents were not affected by fall N rate. Plots receiving fall N also initiated growth earlier in the spring. Optimum N application timing may depend on whether the forage is being hayed or grazed.

[1]Senior Laboratory Technician, UKREC, Princeton, KY (brittanyhendrix@uky.edu); Associate Professor for Forages Extension & Research, UKREC, Princeton, KY (chris.teutsch@uky.edu); Professor for Agronomy/Soils Research & Extension, UKREC, Princeton, KY (jgrove@uky.edu).

EVALUATE THE DIGESTIVE KINETICS OF ALFALFA-BERMUDAGRASS MIXTURES FOR BEEF CATTLE IN THE DEEP SOUTH

M.B. Henson, P. Zessin, S.L. Dillard, M.K. Mullenix, W.B. Smith[1]

In the deep South, one of the hardest times of year for cattle producers is summer due to extreme temperatures and low-quality perennial forages. Incorporating a highly-digestible crop like alfalfa (*Medicago sativa*) into forage mixtures could be economically advantageous; however, there is not much alfalfa produced in the deep South. Our objective was to characterize the digestive kinetics of steers while evaluating the effects of conserved forage type on *in vivo* digestibility of beef cattle systems in the deep South. Three treatments including bermudagrass (*Cynodon dactylon*) hay, alfalfa baleage, and alfalfa-bermudagrass baleage were supplied to four fistulated steers with a calculated daily forage allowance based on intake, body weight, and feed refusals. Each diet was allocated to one of three periods in which all four animals received the same diet due to potential feed spoilage. Each period had a 21-day adaptation period followed by a 5-day collection period to analyze total fecal weight, feed intake, feed refusal, total urine, and rumen fluid. We found significant differences in digestible protein among diets. Alfalfa baleage had 25.8% higher CPD than alfalfa-bermudagrass baleage and alfalfa-bermudagrass baleage had 19.8% higher CPD than bermudagrass hay. When analyzing fiber digestibility, there were only differences between alfalfa baleage and bermudagrass hay. We found that alfalfa baleage had a 17% higher NDFD than bermudagrass hay ($P < 0.02$) and 21% higher ADFD than bermudagrass hay ($P < 0.01$). That said, alfalfa can serve as a nutrient-rich forage additive to diversify feeding strategies and increase digestibility, thus increasing returns in beef producer's operations.

[1]Department of Animal Science, Auburn University, Auburn, AL.

CONTROLLING HIGHLY INVASIVE CHINESE TALLOW (*TRIADICA SEBIFERA*) WITH PRECISION APPLIED HERBICIDES

S.H. Ingram[1]

Chinese Tallow (*Triadica sebifera*), also known as Popcorn Tree, is an invasive, leguminous tree introduced in the southern US from Asia. Cultural and mechanical control methods are labor intensive, and often have little impact. We compared a lower rate of undiluted aminopyralid (Milestone) applied via hack-and-squirt to the labeled rate of diluted Milestone for control of Chinese Tallow while reducing total product applied and number of hacks per tree. A field trial utilizing trees averaging 4-inch diameter at breast height (DBH) examined efficacy of, 1) Milestone at 10% V/V of 1.0 mL, 2) Milestone at 50% V/V of 1.0 mL, 3) Milestone at 100% V/V of 0.5 mL, 4) Aminocyclopyrachlor at 100% V/V of 0.5 mL, and 5) Untreated hack. All cuts on trees were made at a 45°angle at 3-4 feet above soil surface using a machete. Treatment 1 followed label instructions (overlapping hacks with diluted solution applied in every hack), and 2, 3, and 4 were applied to one hack per 4 inch DBH. Herbicide was applied immediately after the cut using a continuous injector syringe. Number of hacks averaged 8.5, 2.5, 3.2, 2.8, and 2.5 for treatments 1,2,3,4 and 5, respectively. Average amount of solution applied was 8.5, 2.5, 1.6, and 1.4 for treatments 1,2,3, and 4, respectively. A visual rating of percentage leaf brownout (Leaf B/O) for each tree was taken 30 days after application to give an early indication of herbicide efficacy. Leaf B/O averaged 90, 87, 90, 96, and 0 for 1,2,3,4 and 5, respectively. These data indicate there is potential to control Chinese Tallow while reducing total solution applied and total hacks made in comparison to the current label recommendations. Pursuance of Special Local Needs (SLN) labels may be an option for states with invasive Chinese Tallow.

[1]Field Scientist, Corteva agriscience, Savannah, GA (sam.ingram@corteva.com)

IDENTIFYING DISEASE IN GEORGIA ALFALFA

F.B. Jackson, J.L. Williams-Woodward, and J.K. Fielder[1]

Alfalfa (*Medicago sativa*) production is economically and nutritionally important to livestock and equine, but the management of alfalfa is more intensive than other forages. Georgia soils are typically lower in pH, and alfalfa requires a minimum of 6.5 pH in order to establish, grow, and promote nutrient intake from phosphorous and potassium applications. Soils with lower pH are at risk of aluminum toxicity. However, with Georgia's higher humidity level than much of the rest of alfalfa production areas, disease is certainly a major concern. The three most popular varieties planted in Georgia are Bulldog 505, Bulldog 805, and Alfagraze 600RR. Bulldog 505 and Bulldog 805 exhibit resistance to Fusarium wilt and Phytophthora root rot. Alfagraze 600 RR exhibits resistance to nematodes, Phytophthora root rot, Fusarium wilt, Anthracnose, and Verticillium wilt. Disease resistant varieties do alleviate some management to several major pathogens known to infect alfalfa. Three sites in Jones and Putnam County were scouted from September 2022 to March 2023. During this time, four diseases were observed and confirmed through microscopy which included common leaf spot (*Pseudopeziza medicaginis*), Stemphylium leaf spot (*Stemphylium spp.), Rhizoctonia solani*, and Leptosphaerulina leaf spot (*Leptosphaerulina briosiana*). Additionally, of the 1,339 plants evaluated during this period, 84% exhibited signs or symptoms of disease. Future research would include a comprehensive study of diseases in relation to alfalfa in Georgia in order to determine the feasibility or economic impact of control through chemical applications.

[1]County Extension Agent, University of Georgia, Gray, GA (brennan4@uga.edu); Associate Professor & Plant Pathologist, University of Georgia, Athens, GA (jwoodwar@uga.edu); County Extension Agent, University of Georgia, Eatonton, GA (kfielder@uga.edu).

RED CLOVER MATURITY AND CULTIVAR AFFECT FORAGE ISOFLAVONE LEVELS

A.A. Jacobs[1], G. Olson[2], J.B. May[2], S.R. Smith[2], T.A. Hamilton[1], and B.E. Davis[1]

Red clover forage contains bioactive phytochemicals including isoflavones. Isoflavone supplementation via red clover has been shown to confer growth and vascular benefits to grazing livestock; however, evaluations of agronomic factors that affect red clover isoflavone levels are needed. The study objective was to evaluate isoflavone profiles of different cultivars across the growing season. Six commercially-available red clover cultivars were established in 2019 in field plots in Lexington, KY. In 2022, biomass was harvested over four dates (April – June), measuring cumulative growth. Biomass was freeze dried, ground to 1-mm, extracted, filtered, and enzymatically hydrolyzed prior to laboratory testing using liquid chromatography and mass spectrophotometer instrumentation (LC-MS-MS). Results showed that isoflavone concentrations were affected by harvest time and cultivar ($p < 0.05$), but there were no interactive effects of harvest time and cultivar ($p > 0.05$). Biochanin A (3690 ± 101 µg/g), formononetin (7545 ± 138 µg/g), and genistein (396 ± 16 µg/g) were greatest at the first harvest. Concentrations of all isoflavones, except daidzein, declined throughout the growing season ($p < 0.05$), with biochanin A declining by 35% and formononetin by 42% from the first to last harvests. Averaged over harvest dates, isoflavones differed by 19% (biochanin A) and 16% (formononetin) from the greatest to least cultivars ($p < 0.05$). These results indicate that red clover isoflavones are affected by both plant growth stage and cultivar. In general, results show that maximizing red clover biochanin A concentration can be accomplished by harvesting at vegetative stages.

[1]USDA-ARS Forage Animal Production Research Unit, Lexington, KY, (brittany.davis@usda.gov); [2]University of Kentucky, Lexington, KY.

MACHINE LEARNING-ENHANCED DEVELOPMENT OF SENSOR-BASED PREDICTION MODELS FOR BERMUDAGRASS (*CYNODON DACTYLON* L. PERS.) YIELD IN EASTERN OKLAHOMA

Gabriel Camargo de Campos Jezus, Lucas Freires Abreu,
Brian Arnall, and Alexandre Caldeira Rocateli

For sustainable forage management, it is essential to align herd size with forage availability. Traditional yield assessment techniques, such as clipping and grazing sticks, are laborious and not ideal for managing multiple pastures. This study explores the utility of digital sensors, including laser, ultrasonic, and multispectral, paired with Machine Learning (ML) techniques, to offer an efficient solution. Specifically, the integration of Random Forest and other ML algorithms was investigated for yield prediction in bermudagrass [*Cynodon dactylon* (L.) Pers.] cultivars, given varying nitrogen applications under Oklahoma's environmental conditions. Data preprocessing involved the removal of incomplete records, and a distinct location was reserved for cross-validation. The models, including Random Forest and Multivariate Polynomial Regression (MPR), utilized sensor data, nitrogen rates, and seasonal rainfall for training and testing in an 80-20 split. Although MPR achieved an R^2 of 0.78 during initial evaluations, its performance dropped to 0.34 in the cross-validation, contrasting with Random Forest's consistent R^2 values (0.76 and 0.53). This suggests that the Random Forest model, with its reduced overfitting, offers a more reliable approach for predicting forage yields, potentially transforming forage management practices for improved sustainability and livestock grazing optimization.

"TOXIC THREATS TO LIVESTOCK IN THE MIDWESTERN PASTURE" EXHIBIT

Jacob E. Tower, G.N. Burcham, Jason E. Tower, M.S. Hasenour, and K.D. Johnson[1]

Toxic plants can be a potential threat to livestock wellbeing. A Purdue University multi-disciplinary team developed an outside live-plant exhibit next to the Heeke Animal Disease Diagnostic Laboratory located at the Southern Indiana Purdue Agricultural Center (SIPAC), Dubois, IN. In the fall of 2022, twenty-one different plants, each potentially toxic when consumed by livestock, were selected and individually seeded or transplanted into two-foot diameter corrugated plastic tile rings that were cut to an eight-inch height. A weather-resilient placard with plant common name, Latin name and QR code was attached to a 36-inch stake and placed by the appropriate ring. Information including plant photographs, Latin Name, Identifying Characteristics, Plant Longevity, Where Commonly Found, Time of Most Concern, Compound that Causes Concern, Part of Plant Most Toxic, Livestock Species Affected, if a concern within Pasture, Stored Feed or Both, Clinical Symptoms, Amount Needed to be Consumed for Clinical Symptoms/Death if Known, and References are reviewable when accessing the QR code. Photographs display the full array of plant longevities and different maturity stages. In 2023, the exhibit was visited by International Grassland Congress tourists, Purdue Extension Educators, Grazing School clientele, and by individuals that come to the agricultural center and Animal Disease Diagnostic Laboratory.

[1]Undergraduate Student, Purdue University Horticulture and Landscaping Department, West Lafayette, IN (tower0@purdue.edu); Veterinary Diagnostician, Heeke Animal Disease Diagnostic Laboratory, Purdue University College of Veterinary Medicine, Dubois, IN (gburcham@purdue.edu); Superintendent, Southern Indiana Purdue Agricultural Center, Dubois, IN (towerj@purdue.edu), Purdue University Extension Educator - Crawford County, Leavenworth, IN (scott422@purdue.edu); Professor of Agronomy, Purdue University, West Lafayette, IN (johnsonk@purdue.edu).

EVALUATING PERRENNIAL WARM-SEASON GRASSES FOR FORAGE PRODUCTION IN SEMIARID GRAZING SYSTEMS USING ALTERNATE ROW SEEDING

T.Z. Jones, B. Fowers, and B.A. Mealor[1]

Hay and forage production provide critical support for Wyoming's extensive livestock industry. Fields and pastures used for this purpose are typically dominated by native and introduced cool-season perennial grasses and alfalfa. Historically, rangeland and meadows of northeast Wyoming also included warm-season perennial grasses that matured later and provided crucial late summer and fall grazing resources. A significant challenge in re-establishing warm-season grasses is finding species that effectively compete with highly persistent cool-season grasses currently used in the region. One approach to overcoming competition in these systems is alternate row seeding. We selected 11 perennial warm-season grass varieties and planted each variety alone and in alternate rows with meadow brome, a commonly utilized cool-season perennial grass. Entries include varieties of sideoats grama, little bluestem, big bluestem, sand bluestem, indiangrass, and switchgrass. On-going assessments of establishment, yield, nutritive content, and harvest interval are being conducted on rain-fed and irrigated sites in northeast Wyoming to assess which varieties are best suited for use in mixed-grass semiarid forage systems. Though the evaluation of reintroducing perennial warm-season grasses into semiarid forage production systems is in its early stages, alternate row seeding is demonstrating promise as one method that could be used to achieve forage diversification.

[1]Assistant Manager – Research, Production, & Outreach, University of Wyoming Sheridan Research & Extension Center (tjones38@uwyo.edu); Research Scientist, Asst., University of Wyoming, Department of Plant Sciences (bfowers@uwyo.edu); Director, University of Wyoming Sheridan Research & Extension Center, Professor, University of Wyoming – Department of Plant Sciences (bamealor@uwyo.edu).

STOCKPILED BAHIA – A SECRET SOUTHERN WEAPON AGAINST HIGH PRICES

K. Kelley

Typically forages like fescue or bermudagrass are considered the primary option(s) for stockpiling in Alabama. However, a current Auburn University research project demonstrates that stockpiling bahiagrass is a viable cool season feed option for cattle producers. Researchers looked at tonnage of bahiagrass produced at varying levels of nutrient amendments (nitrogen) added at multiple locations in south Alabama. There was variance in production levels across sites and amendment level. Researchers used a modified version of the Alabama Cooperative Extension System (ACES) bahiagrass grazing budget to estimate costs. All stockpiling repetitions had a cost assignment, even repetitions that received zero amendment additions. The estimated costs associated with the multiple stockpiling repetitions were compared with bahiagrass and bermudagrass hay prices obtained from the ACES Profit Profiles publication. The stockpiled bahiagrass in the study was an economically viable alternative source of cattle feed compared to bahiagrass or bermudagrass hay.

COVER CROP EFFECTS ON SOIL HEALTH

Robert A. Lane and Shyam Nair[1]

The benefits of cover crops for soil conservation include alternate land uses (including grazing) and improving soil organic matter. In recent years, attention has focused on improving soil health by growing cover crops. In this study, Austrian winterpea (*Pisum sativum*), hairy vetch (*Vicia villosa*), daikon radish (*Raphanus sativus* subsp. *longipinnatus*), and berseem clover (*Trifolium alexandrinum* cv. "Frosty") were planted alone and in combination with rye (*Secale cereale* cv. "Elbon"). Rye planted alone was also included in the treatments. A control treatment, in which no cover crop was planted, was also included in the experimental design. However, a heavy crop of volunteer annual ryegrass (*Lolium multiflorum*) in the control treatment plots likely resulted in those plots receiving a significant organic matter contribution from the ryegrass roots. A summer crop of grain sorghum (*Sorghum bicolor*) was planted with a no-till drill each year over the entire experimental area once the cover crops were terminated (late April). Soil samples from each of the cover crop plots were collected to a depth of six inches in early July, 2021. The samples were packaged in airtight containers on ice and sent via overnight mail to Woods End Laboratories to obtain various measures of soil health in which the lab specializes, including soil aggregate stability, Solvita CO_2 respiration, and extractable amino-N (a measure of organic nitrogen). All data were subjected to an analysis of variance to determine treatment effects. No treatment effects stood out as being superior or inferior to the others. Means were not significantly different, thus our results are inconclusive.

[1]Distinguished Professor Emeritus, School of Agricultural Sciences, Sam Houston State University, Huntsville, TX (blane@shsu.edu); Professor, School of Agricultural Sciences, Sam Houston State University (shyam.nair@shsu.edu).

A SURVEY OF HAY PRODUCTION FOR THE EQUINE MARKET

Krista L. Lea[1], Bob Coleman[2], Laurie Lawrence[2], Jen Zimmerman[3] and S. Ray Smith[1]

Alfalfa is a high-quality forage for livestock, including horses, and the sale of alfalfa hay has significant economic impact on the agricultural community. Two Qualtrics surveys were used to gather information on hay buying attitudes, preferences, and practices of horse owners. One survey targeted horse owners while the second survey was directed to individuals who market hay to horse owners, including hay producers and hay brokers. This survey was designed to evaluate the experiences of hay producers and brokers when selling to the horse market. Questions focused on type and quality, price, package size and preservative treatment. A link to this survey was sent to state and national forage and hay associations for circulation among their membership and received over 300 responses. Nearly 40% of hay producers surveyed responded that horse owners feed alfalfa hay because of its high nutritional value. Additionally, 35% of hay producers claimed that more than 75% of their business is equine. The American Horse Council estimates that there are 7.2 million horses in the US, and our study found that more than 85% of horse owners must purchase hay for their animals. In Kentucky and nearby states, the equine market represents significant demand for high quality hay that hay producers would be wise to target.

[1]Dept. of Plant and Soil Sciences, University of Kentucky, Lexington, KY. [2]Dept. of Animal and Food Sciences, University of Kentucky, Lexington, KY. [3]Equine Science and Management undergraduate student, University of Kentucky, Lexington, KY.

GRAZING MANAGEMENT OF COOL-SEASON CRASSES AND LATE-MATURING CLOVERS

R. Lemus[1]

Grazing cool-season annual grasses and late-maturing clovers is a cost-effective method to reduce N application and hay supplementation in the southern USA. The incorporation of small grain species such as cereal rye (*Secale cereale*), oat (*Avena sativa*), triticale (*Secale cereale* × *Triticum aestivum*), and wheat (*Triticum aestivum*) could provide early grazing while annual ryegrass (*Lolium multiflorum*) and clovers can complement later winter and spring grazing. Most cool-season grasses tend to have high nutritive (CP > 15% and TDN > 60%) value from later fall into mid-spring and meet the energy and protein requirements of stocker cattle. Early maturing clovers such as crimson clover (*Trifolium incarnatum*), hairy vetch (*Vicia villosa*), and winter peas (*Pisum sativum*) might limit benefits to the grazing system. The goal is to incorporate later maturing clovers such as balansa (*Trifolium michelianum*) and berseem (*Trifolium alexandrinum*) to compensate for the loss of nutritive value in maturing grasses and maintain a more consistent animal daily gains. Balansa and berseem clovers are also considered low bloating species which makes them ideal for grazing. Inclusion of clovers into grazing systems could also reduce the need for nitrogen application. Several studies conducted at Mississippi State University have indicated that berseem and balansa clovers can offset forage production when compared to grass monocultures fertilized with 100 lb N ac^{-1} during the grazing season. These studies also indicate that annual ryegrass/clover mix can increase pounds of beef per acre compared to small grain/clover systems.

[1]Professor and Extension Forage Specialist, Mississippi State University, MS (Rocky.Lemus@msstate.edu).

EFFICIENCY, YIELD, AND NUTRITIVE VALUE OF PERENNIAL WARM-SEASON GRASS UNDER LIQUID FERTILIZER APPLICATIONS

R. Lemus[1] and J.A. White[2]

One of the most important soil nutrients that impact hay production and crude protein is nitrogen. A two-year study (2022-2023) compared the impact of liquid fertilizer sources to traditional granular fertilizers in forage bahiagrass (*Paspalum notatum*) production. Ten N fertilizer treatments were applied to a well-established 'Tifton 9' bahiagrass. Treatments included a control (C, no N), two granular fertilizers [urea (U, 46-0-0) and urea ammonium sulfate (UAS, 33-0-0)], and four liquid N fertilizers [urea ammonium nitrate solution (UAN, 32%), Royal Grow (RG, 30-0-0), Pasture Booster Prime (PBP, 30-0-0), and Agritech Plus (AGP, 10-20-10)]. RG, PBP, and AGP were applied at the recommended rate of 1 gal/A and to meet 50 lb/N per cut of hay. U, UAS, and UAN were applied at the equivalent of 50 lb/N. There were three N applications and three harvests in both years of the study. Biomass subsamples were analyzed for nutritive value using NIRS. Data were analyzed using harvest frequency as a repeated measure for N treatments. The use of liquid fertilizers at the manufacturer's recommended rate produced results that were generally lower in response to the granular fertilizers (urea and urea ammonium sulfate) and did not improve seasonal forage production compared to the control. When applied at the recommended rate of 50 lb/N for hay production, the liquid fertilizers produced results that were generally intermediate in response. Nutritive value was only affected by harvest date. Cost per ton of dry matter was significantly higher with liquid than with granular fertilizers.

[1]Professor and Extension Forage Specialist (Rocky.Lemus@msstate.edu) and; [2]Forage Variety Testing Manager, Mississippi State University, MS (JWhite@pss.msstate.edu).

GRASS ENDOPHYTES AND MYCOTOXINS IN FLORIDA PASTURES

Hui-Ling (Sunny) Liao, K. Zhang, V. Mendez, H. Wang, Ko-Hsuan Chen, B. Reimer, V. Verma, B. Justesen, J. Walter, A. Blount, C. Mackowiak, M. Wallau, R. Jones, and D. Mayo[1]

Mycotoxins have rendered grass toxicity a significant concern, leading to economic losses in the USA, particularly impacting the health of livestock that graze affected grasses. The changing climate further contributes to proliferation of mycotoxins by grass fungi, including endophytes that reside in plant tissues and epiphytes that reside on plant tissues. Cattle ranchers in Florida have highlighted health issues associated with grazing on grass pastures. Because fungi residing in or on grasses do not exhibit visible symptoms, laboratory testing becomes crucial to assess fungal titers and mycotoxin levels. We currently lack updated information regarding the types of mycotoxins present in Florida pastures under various environmental conditions that hinders ranchers from making informed decisions about management to prevent animals from consuming grasses with elevated mycotoxin levels. To address this, our extension agents collected and examined approximately 300 grass samples from 13 ranchers across Florida over multiple years, encompassing both dry and wet seasons. The grass species included bahiagrass, bermudagrass, limpograss, and smutgrass. DNA sequencing and LC-MS/MS technologies was used to identify and quantify fungal endophytes and mycotoxins present in the collected samples. We found that zearalenone-4-sulfate (*Fusarium* spp.) as the dominant mycotoxin across ranches in Central and South Florida. Dihydro-lysergol (*Aspergillus, Penicillium*) was found in Northern ranches, while Alternariol methyl-ether (*Alternaria spp.*) was the top mycotoxin in samples from Cherokee county. Emodin (produced by grasses or fungi) was identified as a prevalent metabolite in some ranches across the state.

[1]Institute of Food and Agricultural Sciences (IFAS), University of Florida.

ESTIMATING SUPPLEMENTAL FEED COSTS FOR CATTLE BASED ON FORAGE NUTRITIVE VALUES IN TENNESSEE

K.M. Mason, C.N. Boyer, D. McIntosh, B.C. Pedreira, C.C. Martinez[1]

In the Southeastern United States, producers rely on hay to meet cattle nutrient requirements throughout the winter. Prices related to supplemental feeding greatly influence profitability, so it is imperative that hay meets nutritional requirements in order to reduce supplemental feed. To ensure that producers are using the most economical feeding options, supplemental feed costs were estimated for eight species of conserved or utilized forages in Tennessee. Forage testing sample data from Tennessee, 2015 to 2022 (5,148 samples), provided an estimated feed cost/cow/day. Ration recommendations were based on a 14% crude protein (CP) commercial product, with an estimated total digestible nutrient (TDN) of 68%. Average nutritive values for tall fescue and mixed grass hay were 11% CP and 58% TDN. When nutritive value was compared with nutritional requirements for a 1200 lb cow in late-gestation, no supplemental feed was needed where hay intake was not limited. The cost/cow/day for a late-gestation cow averaged $2.43. For lactating cows, no supplemental feed was needed due to adequate nutrient concentrations. However, lactating cows have a greater dry matter intake/day resulting in a cost/cow/day of $2.70. This information is crucial in helping producers decide how to use hay strategically to meet the nutritional requirements of cows based on calving season. Management practices must rely on testing of conserved or utilized forage to determine if feeding requirements are met at each stage of production.

[1]Assistant Professor/Extension Specialist, Department of Animal Science, The University of Tennessee, Knoxville, TN (kmason21@utk.edu); Professor, Department of Agricultural and Resource Economics; Coordinator, UT Beef and Forage Center, Department of Plant Sciences; Associate Professor, Department of Plant Sciences and Director of UT Beef and Forage Center; Assistant Professor in Department of Agricultural and Resource Economics.

EMPOWERING TENNESSEE EXTENSION AGENTS TO TEACH GRAZING SCHOOL PROGRAMS

K.M. Mason, M.E. Fancher, and P.T. Dugger[1]

Equipping county Extension agents with tools, skills, and knowledge is imperative for a strong statewide Extension program. Agents in Tennessee lacked information and teaching aids to educate livestock producers about grazing management practices and utilization of temporary electric fence. Funding from the Sustainable Agriculture Research and Education (SARE) program provided funding to support agent training. A set of four fencing kits containing a variety of step-in posts, polywire, reels, and chargers were assembled for use at Extension programs throughout Tennessee. Three in-service trainings were conducted to discuss the basic concepts of grazing management and temporary fencing. During hands-on demonstrations, agents were given the opportunity to do a pasture walk, assess forage availability, and set up temporary fencing. A total of 43 agents attended the in-service trainings. Program evaluation results indicate that on a scale of 1 to 5, agents rated the overall usefulness of the trainings as a 4.6. Four topics areas were presented, and there was an increase in knowledge for all four topics: (1) purpose of grazing management, +24%, (2) making a grazing plan, +33%, (3) calculating forage availability and stocking rate, +37%, and (4) using temporary fencing kits, +36%. Fencing kits have been used at various county and regional programs in Tennessee and over 750 livestock producers have attended these programs.

[1]Assistant Professor/Extension Specialist, Department of Animal Science, University of Tennessee, Knoxville, TN (kmason21@utk.edu); Graduate Student, Department of Animal Science, University of Tennessee, Knoxville, TN (mfanche4@vols.utk.edu); Program Coordinator, Center for Profitable Agriculture, UT Extension, Columbia, TN (pdugger2@utk.edu).

WHAT I HAVE LEARNED HELPING CLIENTS MANAGE WEEDY GRASSES

M.D. Mauldin[1]

In over 10 years as an Extension Agent, I have helped clients manage their way through all kinds of unwanted vegetation. Almost without exception, the most difficult and discouraging weed management situations I encounter involve the presence and/or increasing prevalence of an undesirable grass species in an improved grass pasture. Taking the effort to quantify why these situations are so challenging can help lead to the development and implementation of more effective strategies to prevent/address these situations. The following are my observations: 1) Weedy grasses are particularly troublesome because they are difficult to distinguish from desirable grasses – they are often not a concern until the problem has become quite severe. 2) Producers understand that issues with several (not all) weedy grasses are strongly tied to poor grazing and/or fertility management. Having to acknowledge this causes some hesitancy in seeking control/management recommendations from sources like Extension. 3) Unfortunately, when an attempt is made to address the problem, the weedy species involved are generally quite hardy and inherently difficult to control. This is exacerbated by the lack of chemical control options available to control the weedy species without severely damaging the desirable species. Control recommendations tend to involve multi-step, multi-year, endeavors that don't guarantee lasting success. Considering the limitations of reactionary responses, Extension should increase its emphasis on preventative measures including, trainings related to grass identification, and "pasture check-ups" to help clients find problems before they become severe. Additional management strategies should also be explored.

[1]Agriculture & Natural Resources Agent, UF/IFAS Extension Washington County, Chipley, FL (mdm83@ufl.edu).

COMPARING HARVEST METHODS OF PEARL MILLET

M. Mobley[1], K.J. Simon[2], and M.S. Gadberry[3]

Pearl millet is a fast-growing warm-season annual grass that requires less nitrogen fertilizer than bermudagrass per ton of forage production. As fertilizer prices have increased producers are looking to alternatives to reduce the dependency on N fertilizer. Pearl millet has become a popular option. However, it presents challenges when harvested as dry hay due to the required field drying time needed to reach the appropriate moisture content. This demonstration aimed to assess the feasibility of various mowing techniques for harvesting pearl millet as dry round bales. Three cutting methods were tested: a flex-wing bushhog, a rotary disc mower, and a rotary disc mower equipped with a flair conditioner. Moisture samples were collected every 24 hours for each harvest method. Drying conditions, including rainfall, temperature, dew point, wind, and humidity were also monitored. The results revealed distinctive characteristics of the mowing techniques. The bush hog method resulted in shredded leaf material, forming a dense flat mat across the ground. Additionally, the bush hogging method caused excessive lodging without cutting stalks. The mower conditioner effectively crushed the plant material to facilitate drying, while the disk mower did not involve any processing. Results indicate that the drying time can be reduced by using a mower conditioner instead of a disc mower. Bush hogging was not a feasible alternative.

[1]Independendence County Agent, University of Arkansas System Division of Agriculture, Cooperative Extension Service, Batesville, AR (mmobley@uada.edu); [2]Forages Instructor, University of Arkansas System Division of Agriculture, Cooperative Extension Service, Little Rock, AR (ksimon@uada.edu); [3]Professor and Resident Director-LFST, University of Arkansas System Division of Agriculture, Arkansas Agricultural Experiment Station, Batesville, AR (sgadberry@uada.edu).

INTERCROPPING ALFALFA DURING ESTABLISHMENT TO ENHANCE CROPPING SYSTEMS

H.M. Mosqueda, M.T. Berti, A. Kurth, F. Omeye, H. Lindell, and Md.S. Islam[1]

Over time, traditional cropping systems will be less productive as high inputs and low diversity alters the soil microbiome and physiochemical properties. Cropping system diversification by including perennial legumes can improve soil and increase yields. However, despite the benefits it can have on soil health, alfalfa (*Medicago sativa* L.) production is declining because its establishment year is often less productive. The objectives of this study are to reduce economic loss in the seeding year and determine whether intercropping corn (*Zea mays* L.) or forage sorghum (*Sorghum bicolor* (L.) Moench) with alfalfa can increase overall yield and nutritive value of the perennial system. Corn (C), sorghum (S), and alfalfa (A), alone or intercropped, (corn (CA) or sorghum (SA) + alfalfa) were planted under different management strategies: traditional and 0.6-m gaps every 1.8-m within corn rows (G, corn gaps; GA, corn gaps + alfalfa) or a two-harvest system of sorghum 45 and 90 days after planting (S2, two-harvest sorghum; SA2, two-harvest sorghum + alfalfa). In Year 1, crude protein was greater in S2 (94.0 g/kg) and SA2 (89.9 g/kg) than in S (71.8 g'kg) and SA (61.2 g/kg). Biomass was similar across all sorghum treatments, and the reduction of corn grain yield in G and GA was less than 17% compared with C and CA. Alfalfa failed to establish under SA, however, it maintained similar stands in SA2 compared with A. Yields and nutritive value for alfalfa in Year 2 are similar across all treatments. Integrating alfalfa into corn and forage sorghum systems can be feasible.

[1]PhD Research Assistant (haley.mosqueda@ndsu.edu); Professor (marisol.berti@ndsu.edu); Graduate Research Assistant (anastasia.kurth@ndsu.edu); Graduate Research Assistant (franklin.omeye@ndsu.edu); Research Specialist (houston.lindell@ndsu.edu) Graduate Research Assistant (mdshazzadul.islam@ndsu.edu), North Dakota State University, Fargo, ND.

EFFECTS OF VARIOUS COVER CROP TERMINATION STRATEGIES ON EARLY WEED SUPPRESSION IN SOYBEANS AND SORGHUM

C.C. Nieman[1], J. Schafer[2], K. Cartwright[3], and L. Ashlock[3]

The incorporation of cover crops and their associated termination methods may help control weeds in organic cropping systems. A study was conducted to compare four organic cropping systems with cover crops and various termination strategies including tillage, crimping, and grazing. System 1 was no-tillage and no grazing (NT); a fall planted annual was crimped in spring prior to no-till planting. System 2 was no-tillage with grazing (NTG); a fall planted annual was grazed in spring prior to no-till planting. System 3 used conventional tillage and grazing (CTG); a fall planted annual was grazed in spring then tilled prior to planting. System 4 was a dual cropping system with conventional tillage (CT); a fall planted annual was harvested for grain, followed by tillage, and planting. Each system had four replicates and plot size ranged from 0.3 to 0.5 acres. The summer cash crop was soybeans in 2022 and sorghum in 2023. Botanical composition was evaluated by a modified Daubenmire method completed four weeks after planting in 2022 and six weeks after planting in 2023. In both years, Johnsongrass was the main weed. In 2022, tilled plots contained more Johnsongrass compared to no-till plots, 20% in CTG and 26% in CT compared to 10% in NT and 3% in NTG. In 2023, botanical composition was not associated with tillage. Johnsongrass was 31% in NT, 13% in NTG, and 23% in CTG, while the later planted CT plots were only 4% Johnsongrass and the main weed was signal grass (14%). Although the main weed among plots was consistent over two years, additional years of research are required to determine differences in weed control among the cropping systems.

[1]USDA-ARS, Dale Bumpers Small Farms Research Center, Booneville, AR; [2]Winrock International, Little Rock, AR; [3]Natural Soybean and Grain Alliance, Fayetteville, AR.

OPTIMIZING BIOGAS PRODUCTION FROM CACTUS (*Opuntia spp.*) FEEDSTOCK

Kenneth T. Oduor[1], José C. B. Dubeux Jr.[1], Willis Owino[2], Marilia A. Bernardini[1], Rafael S. Reis[1], Luana M. D. Queiroz[1], Roger Trump[1], Javier Portuguez[1], Simili[1], Igor L. Bretas[1], Mario A. Lira Jr.[1], Beatriz E. Cremostim[1] and Martin Ruiz-Moreno[1]

Cactus biomass can produce biogas and contribute to the sustainable management of invasive species. We conducted three experiments to evaluate strategies for optimizing biogas from cactus feedstock. First, we sought to co-digest cactus with varying levels of cattle dung (0, 10, 20, and 30%) to identify an optimal amount to enhance biogas' quantity and quality. Adding 20% cattle dung resulted in the highest total gas production at 114.9±0.05 ml, a 43% increase compared to 0%, over a 192-hour incubation period. Second, we evaluated the effect of different sodium bicarbonate levels (0, 0.5, 1 and 1.5%) on biogas production when co-digesting 20% cattle dung with cactus. Typically, cactus has a lower pH than the ideal microbial range of 6.5-7.5. The inclusion of 1.5% bicarbonate adjusted the initial pH to 7.5, resulting in a final pH of 6.12. This pH adjustment enhanced both total gas production and methane yield. Finally, we evaluated how incubating cattle dung (priming) and weekly replacement of cactus with cattle dung influenced pH and methane production. Priming coupled with weekly replacement of cactus with cattle feces enhanced the pH within the optimum range of 6.5 to 7.7 required for the methanogens. The highest cumulative CH_4 102.65±0.11 g was observed in treatment with cattle dung primed for 42 days, along with weekly replacement of cactus with cattle dung. Overall, our strategies enhanced biogas production from cactus feedstock and upscaling their evaluation on a pilot scale is warranted.

[1]University of Florida, North Florida Research and Education Center-Marianna, 3925 Highway 71, Marianna, FL, 3925 USA; [2]Jomo Kenyatta University of Agriculture and Technology, Juja 62000 00200, Kenya.

INCREASED FORAGE YIELD AND NUTRITIVE VALUE ESTABLISHING ALFALFA WITH INTERCROPPING SYSTEM WITH FORAGE SORGHUM

F. Omeye

Alfalfa (*Medicago sativa* L.) is a high crude protein and important forage crop in the world but has low forage yield in the establishment year. This deters farmers from growing alfalfa because it takes up space on farms that could be used for other cash crops. Previous research has shown that new alfalfa stands can be successfully established while growing corn in intercropping without significant differences in forage yield and quality. This research was conducted to evaluate implications of integrating alfalfa in an intercropping system with forage sorghum (*Sorghum bicolor* L.), with the aim of increasing the alfalfa acreage and productivity in the seeding year in ways that are economical and sustainable. This experiment was conducted in Fargo and Prosper, ND, USA in 2023-2024. The design of the experiment was a randomized complete block and four replicates. The treatments for this study included (1) alfalfa alone, (2) alfalfa-sorghum at 40 kg N/ha (40N), (3) alfalfa-sorghum at 80 kg N/ha (80N), (4) sainfoin alone, (5) sainfoin-sorghum at 40N, (6) sainfoin-sorghum at 80N, and (7) sorghum alone. The overall scale of success within these treatments is measured in sorghum yield and impact of alfalfa in the intercropped plots in the first year and alfalfa will be compared with sainfoin yield and general performance in the second year. There was no significant difference in the sorghum forage yield and nutritive value across treatments. Overall high presence of beneficial insects in the plots intercropped with alfalfa supports how alfalfa improves biodiversity. For USA Midwest forage growers, optimizing this advanced cropping system could enhance crop efficiency and productivity, forage nutritional value, as well as profitability.

PRODUCTION OF COOL-SEASON ANNUAL FORAGE MIXTURES IN SOUTHERN VIRGINIA

Kathryn M. Payne[1]

In Southern Virginia, it is common for commodity crop growers to also maintain a livestock herd. With growing interest to pursue year-round grazing systems, an option available to producers is grazing cool-season annual cover crops. This study was conducted to evaluate the production potential and nutritive value of different cool-season annual forage systems. Eighteen treatments of different grass and legume mixtures, including wheat, rye, barley, oats, annual ryegrass, crimson clover, Austrian winter pea, and hairy vetch were included. This small plot study was conducted over three production seasons (2020-2022) at the Southern Piedmont Agricultural Research and Extension Center in Blackstone, VA. In 2022, higher than average temperatures and rainfall resulted in greater total forage yields (5,107 lb DM/A) than in 2020 (3,790 lb DM/A) and 2021 (2,869 lb DM/A). Annual ryegrass and its mixtures evaluated were consistently the highest yielding over the three-year period (average 5,324 lb DM/A) but resulted in the lowest crude protein content (9.3%). Although legume monocultures resulted in the lowest forage yields among treatments, the addition of crimson clover with barley resulted in a 21.7% increase in forage yield and a 16.8% increase in crude protein. Results from this study have shown that including legumes in mixture with cool-season annual grasses can increase forage yields and forage nutritive value.

[1]Applied Forage Systems Specialist, Virginia Polytechnic Institute and State University, Southern Piedmont Agricultural Research and Extension Center, Blackstone, VA (kathrynpayne@vt.edu).

LEGUME ESTABLISHMENT IN NATIVE WARM-SEASON GRASSES

Kathryn M. Payne and Carter B. Phillips[1]

Interseeding legumes into native warm-season grass (NWSG) pastures has been known to be a challenge, but their incorporation may increase the seasonal forage distribution and improve nutritive value available to grazing livestock. The objective of this study was to investigate the effect of planting method, planting timing, and the efficacy of burning residue on legume establishment in NWSG. Two forage legumes, 'Alice' white clover and 'Freedom HR' red clover, were seeded into mixed 'Niagara' big bluestem, 'GA Ecotype' Indiangrass, and 'Camper' little bluestem pasture. Three planting methods (no-till drill, broadcast, and a non-planted control) were evaluated at three planting times (fall planting, winter planting, and winter planting with burned residue) over two years (2022-2023) at the Southern Piedmont AREC in Blackstone, VA. Among the treatment combinations, burned and drilled plots produced more clover plants in spring at 30 plants/ft^2, followed by winter drill (14 plants/ft^2) and burn broadcast (12 plants/ft^2). As a result of greater initial clover emergence, plots that were burned or seeded in the winter had greater clover content throughout the experiment; burned and drilled plots had over 90% clover ground cover throughout the second year. Domination of plots by clover in the second year caused yields and the proportion of NWSG in the stand to decline, with burned plots yielding 5,136 lb DM/A compared to a winter-fall mean of 6,628 lb DM/A. Continued research will evaluate the effect red clover, as a short-lived perennial, has on stand composition over time.

[1]Applied Forage Systems Specialist, Virginia Polytechnic Institute and State University, Southern Piedmont Agricultural Research and Extension Center, Blackstone, VA (kathrynpayne@vt.edu); Graduate Research Assistant, Virginia Polytechnic Institute and State University, School of Plant and Environmental Sciences, Blackstone, VA (carterp00@vt.edu).

EVALUATING BALE GRAZING AS A TOOL FOR INCREASING SOIL FERTILITY IN PASTURES

W.E. Peck, B.L. Hendrix, C.D. Teutsch, and K.R. VanValin[1]

In the southeastern United States, pasture productivity is commonly limited by low soil fertility. High fertilizer prices often limit applications in low input pastoral systems. An alternative to conventional fertilizer may be using bale grazing to distribute fertilizer nutrients contained in hay. In this system, bales are uniformly distributed throughout the pasture prior to introducing cattle, and are allocated to cattle using temporary electric fencing. Little data are available for bale grazing in the southern U.S. Our objective was to document the impact of bale grazing on soil fertility. This project was conducted at the University of Kentucky Research and Education Center in Princeton, KY. Forty bales with an average weight of 822 lb/bale were uniformly distributed in a 10 A pasture. The pasture was stocked with 61 cow-calf pairs on 07-Mar-23 and animals were removed on 20-Mar-23. Hay rings were used to minimize feeding losses, and temporary electric fencing was utilized to allocate enough hay for a 2 to 3-day period. Initial soil nutrient levels were documented by sampling every 20 ft along two transects that transversed the pasture. Results indicated that soil pH was in the optimal range, but soil P and K were in the low range. Average residual hay following bale grazing was 89 lb DM/bale with an average area of disturbance of 682 sq ft/bale. Average manure pile density was 1,513 piles/A with an average dry weight of 0.70 lb/pile. In a post grazing assessment, 65% of the pasture area showed disturbance with an average pugging depth of 1.1-in. This project will be repeated to document cumulative impacts of bale grazing over time.

[1]Undergraduate Intern, UKREC, Princeton, KY (Whitney.Peck@uky.edu); Laboratory Technician Senior, UKREC, Princeton, KY (brittanyhendrix@uky.edu); Associate Extension Professor and Forage Specialist, UKREC, Princeton, KY (Chris.Teutsch@uky.edu); Assistant Extension Professor and Beef Specialist, UKREC, Princeton, KY (Katie.VanValin@uky.edu).

GRAZING MANAGEMENT AND ITS IMPACT ON SOIL HEALTH IN THE SOUTHEASTERN USA

M.H. Poore[1] and A.J. Franzluebbers[2]

Pastures are a key component of sustainable ruminant livestock production in the southeastern USA. Grazing of forages can be on perennial pastures, annually planted species for high quality and pasture renovation, or even crop residues or cover crops on cropland. Managing the appropriate class, stock density, and timing of grazing should be tailored to meet the needs of the grazing animal. Grazing management style may have a significant impact on soil health, primarily through the vigor of the botanical sward, return of excreta to the pasture, and residual forage mass that covers the soil. Amendments to feed the sward and removal of forage in hay can also significantly impact soil health. Inappropriate stocking density and/or rate may lead to compromised forage stands that can spiral into poor soil health and high cost of pasture management. This presentation focuses on a variety of grazing management strategies that has relevance for assessment of soil health. Soil health will be described using several important soil characteristics, such as soil organic carbon, nutrient elements from routine chemical analysis, soil aggregation, and soil-test biological activity.

[1]Animal Scientist, North Carolina State University, Raleigh NC 27695 (mhpoore@ncsu.edu)
[2]Reseach Ecologist, USDA-Agricultural Research Service, NCSU Campus Box 7620, Raleigh NC 27695 (alan.franzluebbers@usda.gov).

TRACKING NITROGEN PATHWAYS ON CONTRASTING CROPPING SYSTEMS UTILIZING ^{15}N-LABELED-UREA

J. Portuguez Acuña, J. C.B. Dubeux Jr., L. Sollenberger, C. Mackowiak, N. Dilorenzo, A. Martin-Ryals, K. Roger Trumpp, L. Queiroz, K. Oduor, I. Lima Bretas, M. Araujo Bernardini, M. Ruiz-Moreno, and L. Garcia[1]

Cover crops play a pivotal role in enhancing the sustainability of agricultural systems by providing a range of ecosystem services. Integrating livestock into cover cropping systems offers numerous benefits including improved nutrient cycling and reduced environmental impact such as nitrate leaching. However, the existing body of research on integrated crop-livestock systems lacks comprehensive investigations into N pathways. Our study aimed to track diverse N pathways by utilizing ^{15}N-labeled urea within distinct cropping systems. The investigation was carried out at the North Florida Research and Education Center in Marianna, Florida. Five cover cropping systems were analyzed, i) No cover crop (Control, ii) Oat cover crop + 34 kg N/ha and no grazing, iii) Sod based rotation (SBR), oat/crimson clover + grazing iv) SBR oat/crimson + 34 kg N/ha + grazing, v) SBR oat/crimson clover + 90 kg N/ha + grazing. Emission factors and cumulative emissions were analyzed for nitrous oxide emissions and N15 on soil, feces, and urine over 28 days. The results show that urine was the greatest emission factor followed by feces and soil, but total cumulative emissions were greater in feces than urine and soil.

[1]MS. Student (portuguez.javier@ufl.edu); Professor Forage Agronomy (dubeux@ufl.edu); MS. Student (ktrumpp@ufl.edu); PhD Student (ldantasqueiroz@ufl.edu); Post-doctoral associate (ig.limabretas@ufl.edu); PhD Student (kenneth.oduor@ufl.edu); MS Student (m.araujobernardi@ufl.edu); Biological Scientist (ruizm001@ufl.edu); Research Coordinator (lizagarcia@ufl.edu) - North Florida Research and Education Center, Marianna, FL; Professor Forage Agronomy Department, Gainesville, FL (lesollen@ufl.edu); Associate Professor Soil and Water Science, North Florida Research and Education Center, Quincy, FL (echo13@ufl.edu); Professor Animal Nutrition, North Florida Research and Education Center (ndilorezno@ufl.edu); Professor Agricultural and Biological Engineering, Gainesville, FL (admartin@ufl.edu).

TAMING FESCUE TOXICOSIS: EFFECTS OF SERICEA LESPEDEZA SUPPLEMENTATION ON STEERS

S. Poudel[1], G.J Pent[2], J.H. Fike[1], B. Davis[3], and W. Zeller[4]

Fescue toxicosis poses a significant challenge to the United States beef industry, resulting in considerable economic losses. While condensed tannins (CT's) in specific leguminous forages have the potential to reduce the absorption of toxic alkaloids, their use as supplements is relatively unexplored. This study investigated the potential of sericea lespedeza (*Lespedeza cuneata*) pellets, a CT-containing feed supplement, in mitigating the post-ingestive effects of fescue toxicosis. In 2021 and 2022, twelve steers grazing toxic fescue were supplemented with sericea lespedeza (LES) or LES mixed with polyethylene glycol (LPEG, positive control) for 12 weeks. The study assessed steer body weight, hair retention score (HRS), rectal and extremity temperatures, caudal artery lumen area, and cortisol levels. Steers on the LES diet had lower HRS compared to those on the LPEG diet across both years. In 2021, steers on the LES diet had a greater ADG, but in 2022, ADG did not differ between treatments. LES-fed steers showed hotter tail skin temperatures, cooler rectal temperatures, and a 21% larger caudal artery lumen area than steers fed LPEG, indicating reduced vasoconstriction. LES-fed steers exhibited lower hair cortisol levels, suggesting reduced chronic stress. These findings suggest that CT may ease fescue toxicosis effects, but their potential to reduce alkaloid absorption remains unanswered. Further research is needed to elucidate underlying mechanisms and establish the long-term efficacy of such interventions.

[1]Postdoc, Virginia Tech, Blacksburg, VA (sanjokp@vt.edu); Professor, Virginia Tech, Blacksburg, VA (jfike@vt.edu); [2]Superintendent, SVAREC, Virginia Tech, Raphine, VA (gpent@vt.edu); [3]Animal Scientist, USDA-ARS, Forage-animal Production Research, Lexington, KY (brittany.harlow@usda.gov); [4]Research Chemist, USDA-ARS, U.S. Dairy Forage Research Center, Madison, WI (wayne.zeller@usda.gov).

ASSESSING VARIABLE RATE SEEDING TECHNOLOGY FOR LOW INPUT PASTORAL SYSTEMS

C.G. Roper, B.L. Hendrix, and C.D. Teutsch[1]

Abstract

Frost seeding is the practice of broadcasting seed in late winter and allowing freezing and thawing cycles to incorporate the seed into the soil. It is commonly used to reintroduce clover back into pastures. The rough and uneven terrain often found in pastures is not conducive to maintaining a consistent speed when frost seeding. This results in the over and under application of seed. Recently a company out of Austria has developed a spinner type seeder that is capable of varying seeding rate in real-time as the speed of the UTV changes. The objective of this study was to evaluate the impact of utilizing variable seeding rate technology on actual seeding rate as ground speed was varied. There were six treatments in the study: ground speeds of 3, 6, and 9 mph with variable or constant seeding rates. When ground speed was increased from 6 to 9 mph and a constant seeding rate was used, the actual seeding rate decreased by 30% and when ground speed was decreased from 6 to 3 mph, the actual seeding rate almost doubled. When the variable seeding rate option was used, the actual seeding rate remained relatively constant (< 10% variation) regardless of changes in ground speed. Overall, using variable seeding rate technology could reduce seeding costs for livestock producers and result in more uniform stands of clover. However, the high cost of this technology may limit adoption.

Introduction: Frost seeding is the process of broadcasting clover or other legume seed onto existing pastures or hayfields in late winter and allowing freezing and thawing cycles to incorporate the seed into the soil (Teutsch et al., 2021). This method is most commonly used with red and white clover as well as annual lespedeza, all of which are legumes and an essential part of sustainable grassland ecosystems (Whitehead, 2000).

Variable rate seeding (VRS) technology allows seeding rate to be adjusted in real time as field conditions or ground speed changes (Šarauskis et al., 2022). This could be especially useful in pastures where rough terrain makes maintaining a consistent speed difficult (Fig. 1). This technology could improve the uniformity of seed dispersal resulting in more uniform stands of clover. The objective of this study was to evaluate the impact of variable seeding technology on actual seeding rate as ground speed was varied.

Materials and Methods: This experiment was conducted at the University of Kentucky Research and Education in Princeton, KY. The experimental design was a random complete block with four replications. An UTV (Kawasaki Mule, 4010) was driven on a paved course that was 1,000 ft in length (Fig. 2). An APV broadcast seeder, model MDD 100 M1 (APV America Inc., Pottsboro, TX), was mounted in the bed of the UTV. The seeder was calibrated for a seeding rate

[1]Master Grazer Coordinator, UKREC, Princeton, KY (caroline.roper@uky.edu); Laboratory Technician Senior, UKREC, Princeton, KY (brittanyhendrix@uky.edu; Extension Associate Professor, UKREC, Princeton, KY (chris.teutsch@uky.edu)

of 10 lb/A at 6 mph. The treatments consisted of driving the course at 3, 6, and 9 miles per hour (mph) with either variable or constant seeding rates. A Raven CR7 GPS unit (Raven Applied Technology, Sioux Falls, SD) was used to monitor speed. The speed was also calculated by timing each run and converting that measurement into mph.

At the beginning of each treatment, 15 pounds of wheat seed was weighed and added to the seeder. Variable or constant seeding rate was then selected based on the treatment. The course was then driven at the speed designated for that treatment. At the end of the course the seed metering mechanism was disengaged, and the remaining seed was removed with a shop vacuum and weighed (Figures 3 and 4).

Data were analyzed using the General Linear Model Procedure from SAS (SAS Institute, Cary, NC). Means were separated using Fisher's protected least significant difference. Differences were considered significant at $P \leq 0.05$.

Results and Discussion: Actual speeds were not different for the constant and variable seeding rate treatments for a given speed treatment ($P > 0.05$; Fig. 1). In addition, actual speeds for the variable or constant seeding rates were very close to the target speeds of 3, 6, and 9 mph (Fig. 1).

As expected, when the seeding rate remained constant and the speed was varied, actual seeding rates were different (Fig. 2). Increasing speed from 6 to 9 mph resulted in a 30% reduction in the seeding rate. In contrast, slowing ground speed from 6 to 3 mph doubled the seeding rate. When the variable seeding rate option was used, the seeding rate remained relatively constant (<10% variation) regardless of changes in ground speed (Fig. 2).

Summary and Implications: Using variable rate seeding technology in low input pastoral systems could have a measurable impact on pounds of seed applied per acre, especially in uneven terrain. Overall, using variable seeding rate technology could reduce seeding costs for livestock producers and result in more uniform stands of clover. However, cost of this technology will likely limit adoption. The cost of the seeder used in the study was $5,800. This a substantial increase when compared to less automated seeders that cost $500 to $1,000. One potential path to adoption is for Extension offices and Soil and Water Conservations Districts to purchase seeders and make them available as part of already existing equipment share/rental programs. Future research should focus on field experiments that better document the economic impact of utilizing variable seeding rate technology in low input pastoral systems.

Literature Cited:

Šarauskis, E. M. Kazlauskas, V. Naujokienė I. Bručienė, D. Steponavičius, K. Romaneckas A. Jasinskas. 2022. Variable Rate Seeding in Precision Agriculture: Recent Advances and Future Perspectives. Agriculture 2022, 12(2), 305; https://doi.org/10.3390/agriculture12020305.

Teutsch, C.D., S.R. Smith, and J. Henning. 2021. Frost seeding clover: A recipe for success. AGR-271, University of Kentucky. Available at http://www2.ca.uky.edu/agcomm/pubs/AGR/AGR271/AGR271.pdf.

Whitehead, D.C. 2000. NUTRIENT ELEMENTS IN GRASSLAND, Soil–Plant–Animal Relationships. CABI Publishing, New York, NY. Available at http://sherekashmir.informaticspublishing.com/381/1/9780851994376.pdf.

Figure 1. Pastures often contain rough and undulating terrain that makes maintaining a constant speed difficult.

Figure 2. A UTV with the APV seeder mounted in the bed was driven on a paved course that was 1,000 ft in length.

Figure 3. Sweeping remaining seed out to determine the seeding rate by difference.

Figure 4. Weighing remaining seed to determine the seeding rate by difference.

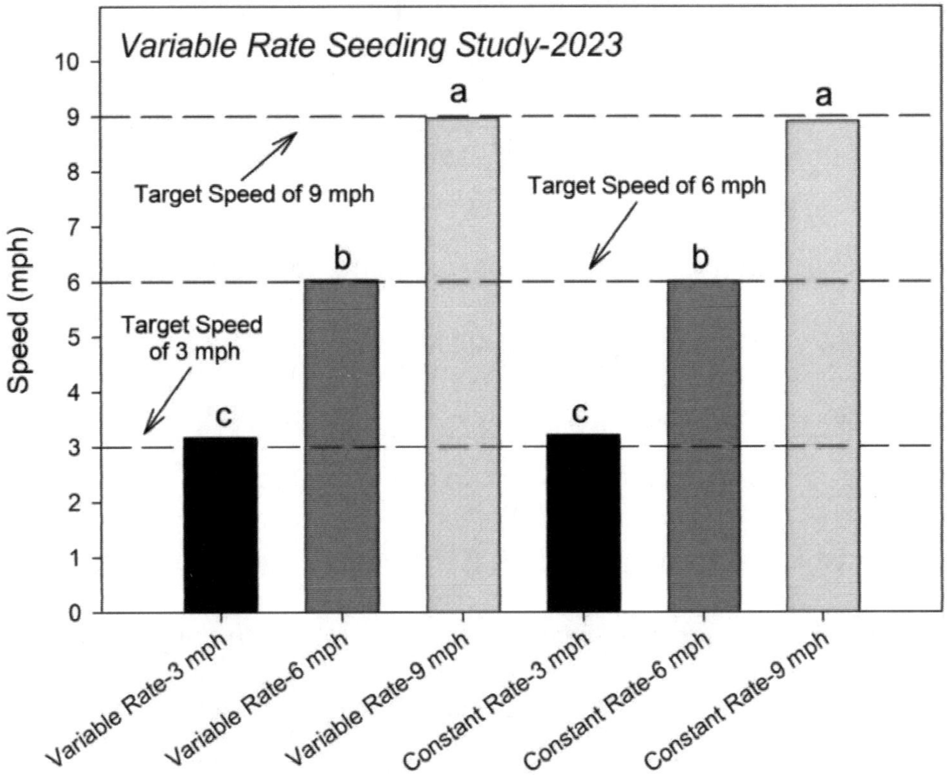

Figure 5. Actual speed in miles per hour for the 3, 6, and 9 mph treatments.

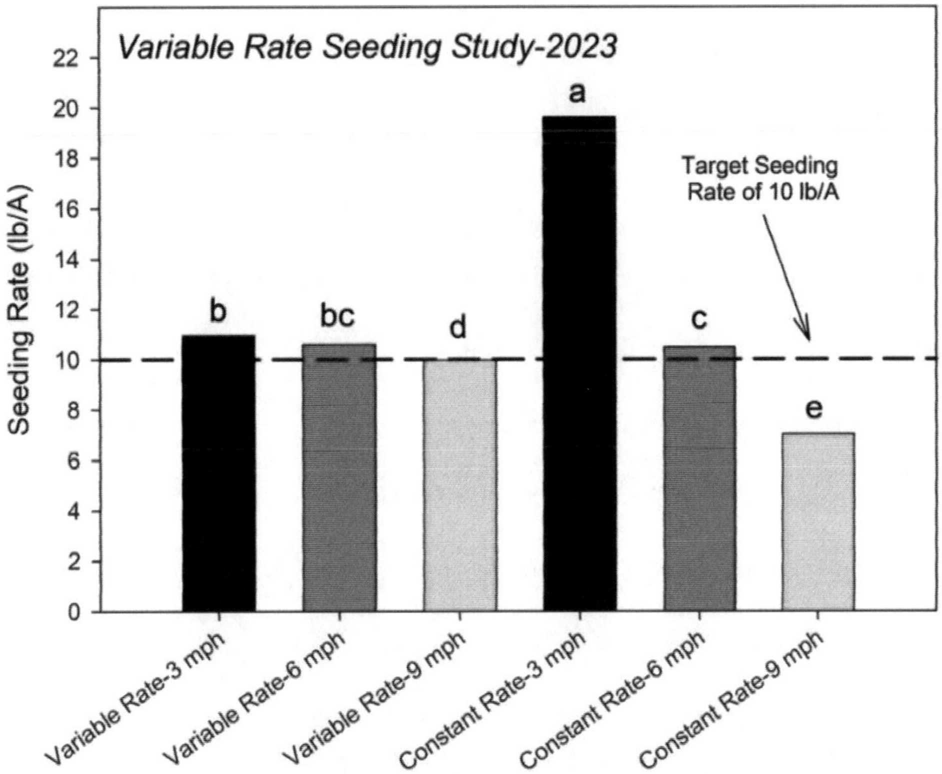

Figure 6. Actual seeding rate in pounds per acre.

PRODUCTIVITY AND NUTRITIVE VALUE OF WARM-SEASON ANNUAL LEGUMES IN MISSISSIPPI

M. Ross[1] and R. Lemus[2]

Warm-season annual legumes could allow producers the opportunity to complement the less nutritive forage crops such as bahiagrass and bermudagrass. The objective of the study was to evaluate the production potential and nutritive value of non-traditional AWSLs in a single cut. The experimental design was a randomized complete block design replicated four times. Four cowpeas' cultivars ('Ace,' 'Common,' 'Purple Hull,' and 'Red Riper') and 'Tyrone' forage soybean were planted 100 lb PLS while aeschynomenee (deer vetch), alyce clover, hairy indigo, sesbania, and lab lab were planted at 25, 20, 15, 40, and 25 lb PLS ac^{-1}, respectively. Plots (6 ft x 10 ft) were established at two planting dates (May 25 and June 28, 2023) and harvested on July 31 and August 21, 2023, respectively. Each species was fertilized with P_2O_5 and K_20 according to soil test recommendations. Before harvest, each species was evaluated for visual botanical composition in a legume percentage basis. Species were harvested using a Wintersteiger harvester (Wintersteiger AG, Austria) when 50 % of the plots reached a height of 12 inches. Biomass subsamples were dried at 140 °F for 72-h and ground to pass a 1-mm screen. Samples were analyzed for nutritive value using the 2023 legume grass equation from the NIRS Forage and Feed Testing Consortium (Berea, KY). Biomass production and legume composition were affected by planting date and species ($P = 0.0045$ and $P < 0.0001$), respectively. Nutritive value (CP, ADF, NDF, IVTDMD, and WSC) were affected by legume species. Crude protein values range from 15.8 to 26.9 DM percentage.

[1]Undergraduate Apprentice, Department of Agricultural Economics, Mississippi State University, MS (mtr110@msstate.edu); [2]Professor and Extension Forage Specialist, Mississippi State University, MS (Rocky.Lemus@msstate.edu).

INTEGRATED CROP-LIVESTOCK SYSTEMS IN MISSISSIPPI

B. Rushing[1], F. Davis[2], and J.C. Lyles[3]

Integrated crop-livestock systems (ICLS) are designed to produce both cash crops and livestock simultaneously on the same land base, which can reduce nutrient concentration and environmental risk, improve soil structure and function, exploit natural pest control processes, and increase overall farm productivity and economic sustainability. A field trial was conducted at the Coastal Plain Branch Experiment Station (CPBES) in Newton, MS and the Prairie Research Unit (PRU) in Prairie, MS from 2021-2023. Soils consisted of a Boswell sandy loam and a Houston clay at CPBES and PRU, respectively. The trial consisted of a randomized complete block design with three replications. Treatments included conventional soybean (CS); un-grazed cereal rye cover crop followed by no-till soybeans (CC); and grazed cereal rye cover crop followed by no-till soybeans (GC). Cover crop production (FM, CP, and TDN), soybean grain yield, animal performance (average daily gain, ADG; gain per acre; GAIN) and economic comparisons were compared for each treatment. Mean cover crop FM (1266 vs. 936 lb DM/A), CP (23.4 vs. 12.6%), and TDN (54.8 vs. 53.3%) were greater at CPBES compared to PRU, resulting in more grazing days (51 vs.10), GAIN (306 vs. 10), and ADG (3.22 vs. 1.16 lb/hd/A). Subsequent soybean yield was unaffected by treatment at CPBES, while the addition of cover crops (CC) and grazing (GC) resulted in 11 and 22 bu/A loss in yield, respectively. These results show the influence soil texture has on the capacity for ICLS to perform in Mississippi. Complete economic and soil analysis for this field trial is pending.

[1]Associate Research/Extension Professor; Mississippi State University; Coastal Plain Branch Experiment Station, Newton, MS; [2]Research Assistant I; Auburn University; Department of Crop, Soil, and Environmental Sciences, Auburn, AL; [3]Research Associate II; Mississippi State University; Coastal Plain Branch Experiment Station, Newton, MS.

PROTOCOL FOR EVALUATION OF FORAGE GRASS MATURITY

S. Ray Smith, Kim Cassida, Roeland Kapsenberg, Jerry Hall, Robin Lamp, Jerome Magnuson, Trent Tate, Devesh Singh, Steve Reid, Krista Lea, James Dedecker, Gabriel Roberts[1]

Abstract

There is a wide range of reproductive maturities among forage grass species and varieties sold in the U.S., but no uniform rating system for producers to use when selecting varieties. To address this issue, seed industry and university representatives formed the AFGC Forage Grass Maturity Working Group in 2016. This group developed a protocol to evaluate and assign varieties into a series of nine maturity groups, similar to the long-standing fall dormancy rating system for alfalfa varieties. This paper outlines how varieties should be evaluated when assigning them into maturity groups.

Introduction: Cool-season grasses can exhibit considerable variation in heading date within a species, but there is no uniform rating system for producers to use when selecting varieties. Previous research has focused on maturity ratings within a single species like the regional orchardgrass maturity comparison by Robins et al. (2017) and the maturity rating system used by Olson et al. (2021) for the University of Kentucky variety testing program. The objective of this research was to develop a rating system that allowed cool season grass varieties to be assigned into maturity groups relative to each other, similar to the alfalfa fall dormancy rating system (Tueber et al., 1998). The grass maturity rating system was developed using Kentucky bluegrass, orchardgrass, tall fescue, bromegrasses, annual ryegrass, hybrid ryegrass, perennial ryegrass, and timothy and is suitable for rating these species and other cool season grass species that mature within the same timeframe.

Materials and Methods: The protocol revolves around the establishment of replicated plots of individual space plants and then monitoring the heading dates of these plants. Individual plants are established under greenhouse conditions as follows: 1) Containers - cell trays that facilitate good root development and ease for transplanting; 2) Planting media - topsoil or potting mix; 3) Temperature and lighting in the greenhouse - 68 to 78°F; 16+ hour photoperiod; 4) number of plants - 25 to 40 per variety per replication, insure only one plant per cell; 5) number of replications – minimum of three; and 6) Other greenhouse management - spray and fertilize as necessary to maintain vigorous growth.

Establishment in the field following these guidelines: 1) Size for transplanting - when plants are approximately 8 to 10 weeks old (when plants can be pulled from tray and roots hold the plug together); 2) Placement in the field - transplant individual plants to the field (late August to early September in northeast and northcentral USA; late September to early October west coast and southern USA); 3) Suggested spacing - 1.5 ft between plants in a row and 2.5 ft row

[1]University of Kentucky (raysmith1@ukyedu); Michigan State University; Jethro Ag Services; GO Seed; Barenbrug USA; DLF USA; GO Seed; Wilbur Elis; DLF USA; University of Kentucky; Michigan State University; and University of Kentucky, respectively.

width as suggested by Plant Variety Protection (PVP) and the grass variety review board (adjust row spacing as needed based on weed control method between rows); 4) Replications - minimum 3 reps and minimum of 15 plants per variety per replication; and 5) Field management – keep nursery as weed free as possible.

Management of plants in the field should be as follows: 1) Clipping - plots may be clipped as needed one to two times during the fall to encourage tillering or reduce overwintering dead residue with the last clipping date based on local climate and no clipping in the spring; 2) Fertilization - based on plant requirements, essential that all plots be fertilized equally; and 3) Weed control - maintain as weed free as possible through tilling, herbicides or other methods, but avoid the use of weed matting or mulch as these may influence maturity date.

Rating plants for maturity date should be as follows: 1) Assigned date - record heading dates as day of year (Jan. 1 is day one, Feb. 1 is day 32, etc…); 2) Maturity rating - assign heading date to individual plants within each plot as the first day when there are at least three fully emerged panicles per plant (ie.-fully emerged from the collar); 3) Plot maturity score - maturity date for the entire plot is recorded as the average heading date across all plants within plot and important to take ratings at least three times per week to ensure precision of measurement; 4) Years - ratings should occur over two growing seasons including the first spring/early summer following fall transplanting (year 1) and the second spring/early summer after transplanting (year 2); 5) Length of trial - rating in the third growing season is not recommended because the research that developed these maturity groups was only based on year 1 and year 2 data; and 6) Growing degree day (GDD) - growing degree days and soil temperature may be collected for future documentation.

Results: The following check varieties will be used to assign a grass maturity group rating for the variety or varieties of interest: Balin – Kentucky bluegrass, Quickdraw – Orchardgrass, Kora – tall fescue, Bariane – tall fescue, Barsprinter – perennial ryegrass, Remington – perennial ryegrass, Polim – perennial ryegrass, and Barpenta – timothy.

It is important to note that not all cool season species have check varieties represented in the standard grass maturity test since many species/variety combinations did not show a consistent maturity ranking over locations and years. The list above shows the varieties that had consistent ranking over years and locations (Figure 1). This single set of check varieties is used to designated grass maturity groups (GMG) 1 to 9. These check varieties have been selected to differentiate the range of relative maturities for cool season grasses using data from an extensive multi-location trial (2015-2023) with 70 varieties from 8 species (Cassida et al., 2023).

The assignment of new varieties into cool season grass maturity groups is determined using the following maturity group rating system.

Maturity Group	**Maturity Group Rating**
1	earlier or equal to Balin
2	later than Balin and less than or equal to Quickdraw
3	later than Quickdraw and less than or equal to Kora
4	later than Kora and less than or equal to Bariane
5	later than Bariane and less than or equal to Barsprinter

6	later than Barsprinter and less than or equal to Remington
7	later than Remington and less than or equal to Polim
8	later than Polim and less than or equal to Barpenta
9	Later than Barpenta

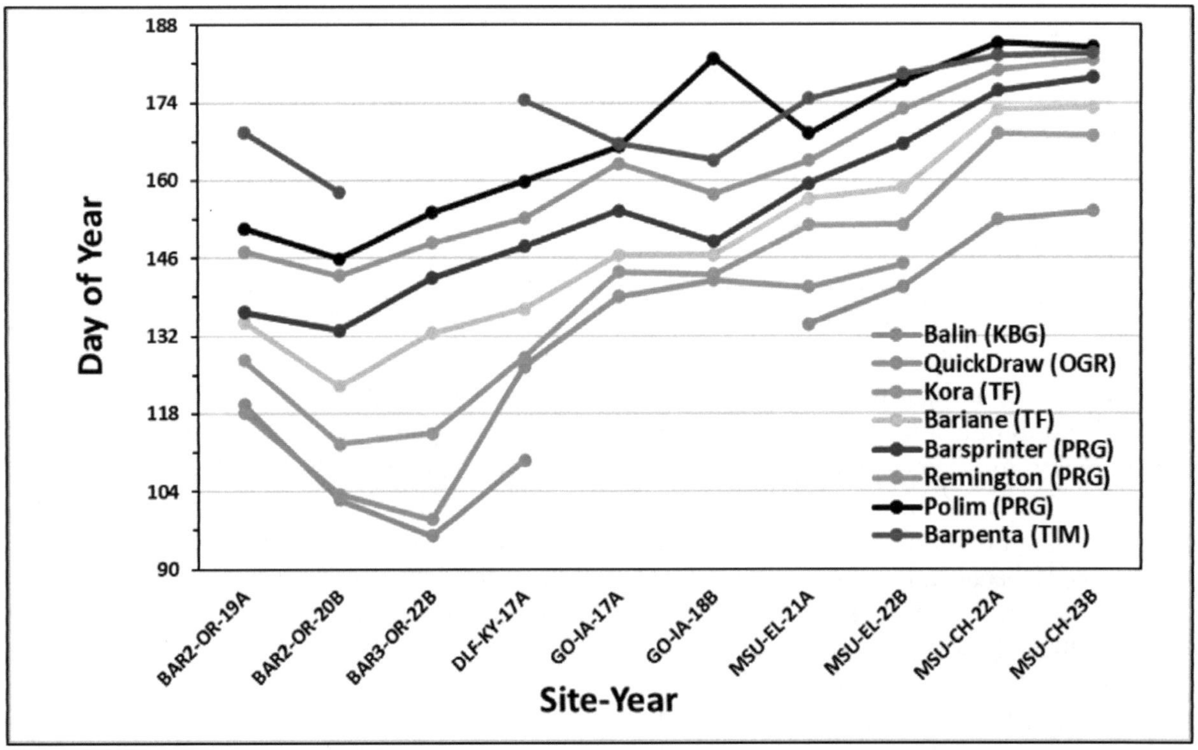

Figure 1. Cool season grass maturity heading dates shown over test locations. These varieties showed consistent ranking over locations and years and were used to develop the list of designated check varieties. Heading date is based on days from January 1 and locations included Oregon (OR), Kentucky (KY), Iowa (IA), and Michigan (MSU).

<u>Discussion</u>: There are numerous considerations when comparing test results over years. Although cool season grass maturity is a strongly expressed trait, genotype by environment interactions do exist based on preliminary tests by Michigan State University. Accurate separation of varieties and assignment of grass maturity group ratings are best obtained by testing for two years at three locations. However, acceptable ratings can be obtained from a single location with year 1 and year 2 ratings for maturity, if the test is conducted at a site representative of the variety's area of adaptation.

Interested parties can make requests for seed of the check varieties using the following procedure. The University of Kentucky will maintain and distribute seed of check varieties from 2024-2028. Upon written request, 10g of seed of each of the check varieties will be distributed to USA-based seed companies and Universities. Send your request to both Dr. Ray Smith (raysmith1@uky.edu) and Gabriel Roberts (james.roberts@uky.edu). Include name, shipping address, email and phone number with your request.

The American Forage and Grassland Council (AFGC) played a key role in organizing the cool season grass maturity committee and will continue to be involved. AFGC will 1) Maintain and publish a list of forage grass check varieties indexed according to the Maturity Rating Protocol; 2) Promote the value and use of the Maturity Rating in grasses; and 3) Maintain a Working Group to monitor usage of the Grass Maturity Rating Protocol; and 4) Update the published list of indexed check varieties on an annual basis. In collaboration with the University of Kentucky over the next five years, the working group will ensure that viable seed is maintained of each check variety. The working group will also ensure that viable seed is maintained in future years by the University of Kentucky or another collaborating University.

This protocol will be sent by the Working Group to the Grass Variety Review board of the American Organization Association of Official Seed Certifying Agencies (AOSCA) https://aosca.org/programs-services/national-variety-review-boards/grass/ and the USDA Plant Variety Protection (PVP) group for acceptance within their systems https://www.ams.usda.gov/services/plant-variety-protection .

Summary: The AFGC Forage Grass Maturity Working Group has developed a protocol to evaluate and assign cool season grass varieties into a series of nine maturity groups. This paper describes the protocol and systematically outlines the management required for plant breeders to implement the protocol from seed to mature plant to the final assignment of maturity groups.

Literature Cited:

Cassida, K., R. Kapsenberg, S.R. Smith, J. Magnusen, S. Reid, R. Lamp, D. Singh, J. Hall, T. Tate, K. Lea, J. Dedecker, and G. Roberts. 2024. Developing a forage grass maturity index. Proc. American Forage & Grassland Conference, Mobile, AL. Jan. 7-10, 2024.

Olson, G.L., S.R. Smith, C. Teutsch, J.C. Henning, and T.D. Phillips. 2021 Orchardgrass report. Univ. of KY Ag. Exp. Station. PR-801. http://www2.ca.uky.edu/agcomm/pubs/PR/PR801/PR801.pdf

Robins, J.G., S.R. Smith, M.H. Hall, C.D. Teutsch, and D.J. Undersander. 2017. Associations among U.S. locations for orchardgrass production. Crop Forage & Turfgrass Management. Vol 3, online Dec 2017. https://doi.org/10.2134/cftm2017.02.0012

Teuber, L.R., K.L. Taggard, L.K. Gibbs, M.H. McCaslin, M.A. Peterson, and D. K. Barnes. 1998. Fall dormancy. North American Alfalfa Improvement Conference. https://www.naaic.org/stdtests/updated/pdfs/FallDormancy.pdf

VALIDATION OF BRIX FOR PREDICTING SUGAR CONCENTRATION OF ALFALFA AND ORCHARDGRASS

E.D. Billman, K.J. Soder, J. Horst, A. Hafla, and K. Balk[1]

Brix has anecdotally been touted as a potential tool for rapid and cost-effective assessment of forage sugar concentrations, but has never been compared to wet chemistry values to verify these claims. This 2-year study evaluated how well Brix values of fresh forage predicted sugar concentration and nutritive value of alfalfa and orchardgrass. Monthly samples of each forage were collected in 2019 and 2021 (May thru Sept) from established swards. Forage samples were immediately evaluated for Brix values with a hand-held digital refractometer, and the remaining biomass was frozen immediately with liquid N to halt cellular respiration. Frozen samples were lyophilized and analyzed for individual sugars and wet chemistry nutritive values. Results showed that Brix values did not reliably equate to corresponding changes in wet chemistry forage nutritive values for either forage. Correlation analyses indicated that Brix was positively correlated (0.357 – 0.516) to individual and total sugars and net energy values in alfalfa and negatively correlated (-0.439 – -0.519) to fiber, ash, and macrominerals (P, K, Ca, Mg). However, Brix values did not correlate to any notable nutritive value parameters in orchardgrass. Findings were attributed to greater fibrous fraction contamination and the much lower sugar concentrations found in grasses and legumes than in horticultural crops. Brix appears to be unreliable at consistently predicting high-sugar forages with desirable nutritive values.

[1]Research Agronomist, USDA-ARS, Coastal Plains, Soil, Water, and Plant Research Center, Florence, SC (Eric.Billman@usda.gov); Research Animal Scientist, USDA-ARS, Pasture Systems and Watershed Management Research Unit, University Park, PA (Kathy.Soder@usda.gov); Laboratory Production Manager, Agri-King, Inc., Fulton, IL (Jeff.Horst@agriking.com;); Beef Nutritionist, Cargill Animal Nutrition, North America, Des Moines, IA (Aimee.Hafla@gmail.com); Assistant Laboratory Production Manager, Agri-King, Inc., Fulton, IL (Kristi.Balk@agriking.com).

EVALUATING AN ALTERNATIVE USE OF FORAGE PRESERVATIVES IN ALFALFA BERMUDAGRASS BALEAGE

B.A. Stefancik[1], J. Pereira[2], P. Zessin[2], L.L. Baxter[3], M.K. Mullenix[2], and J.J. Tucker[1]

Forage preservatives can be a valuable tool when applied at baling, but some Southeastern producers are interested in utilizing preservatives earlier in the process during or immediately following mowing. This application method is thought to decrease dry down time and may improve forage quality. A study utilized the third (June) and fourth (July) cutting of alfalfa-bermudagrass in Tifton, GA, and the fourth (July) cutting in Headland, AL. The objectives were to determine if application of propionic acid preservative at mowing (1) decreased dry down time and (2) impacted nutritive value when harvested as baleage. Treatments were preservative (P+) or no preservative (NP). Forage cores were analyzed for nutritive value in individually wrapped round bales sampled at 6 months post-harvest and (2) mini silos sampled at harvest, 8 weeks, and 6 months post-harvest. P+ did not decrease dry down time to target moisture (65%). Crude Protein (CP), Neutral Detergent Fiber (NDF), or In-vitro True Dry Matter Digestibility (IVDMD) were not different among treatments, locations, or bale package. There was no difference in Acid Detergent Fiber (ADF) at harvest. In mini silos sampled at 8 weeks post fermentation, ADF concentrations were 1.1 and 2.5% higher in P+ forage at both locations, as compared to NP. At 6 months post-fermentation, there was no difference in ADF for either bale package in Tifton, but in Headland, ADF was 2.2% higher in the P+ mini silos. Data suggests that utilizing propionic acid preservative at mowing does not decrease dry down time of forage to target moisture or improve forage quality in alfalfa bermudagrass baleage production.

[1]Animal and Dairy Science Department, University of Georgia, Tifton, GA; [2]Department of Animal Science, Auburn University, Auburn, AL; [3]Crop and Soil Science Department, University of Georgia, Tifton, GA.

EFFECT OF N RATE ON FORAGE YIELD OF LATE SUMMER PLANTED JERRY OAT

S. Stone[1], J. Martin[2], B. Finch[3], K. Simon[4], and D. Kennedy[5]

As temperatures transition from a warm summer to a cool fall/winter, warm-season forage growth slows, and cool-season forage growth increases. This transition period can result in gaps of inadequate forage grazing seasons. Planting winter annuals late in the summer as warm season growth slows can provide ample forage to help fill these gaps in the grazing period. This study was conducted to determine the optimum nitrogen (N) application rate for forage biomass production in late summer planted jerry oats. Jerry oats were planted on September 12, 2023, at 100 lb. seed/A using a Great Plains 606 No-till drill. Nitrogen treatments were applied by hand spreading Urea (46-0-0) at planting at rates of 30, 60, 90, 120, and 150 lb. N/A, along with a no N control to monitor response to N fertilization. To prevent yield limitations and monitor N nutrient impact only, 120 lb. P_2O_5/A and 150 lb. K_2O/A were also applied at planting. Plots were harvested on December 5, 2023, at a 4-inch harvest height by mowing and collecting the biomass. Forage biomass was weighed, and sub-samples were collected for moisture and nutritive value analysis. Application of N resulted in increased forage biomass production of late summer planted jerry oats compared to the control. Unfortunately, this data represents a single year which had an extended dry period following the fertilizer application that reduced the response to applied N.

[1]St. Francis County Agent, University of Arkansas System Division of Agriculture, Cooperative Extension Service, Forrest City, AR (sastone@uada.edu); [2]Cross County Agent, University of Arkansas System Division of Agriculture, Cooperative Extension Service, Wynne, AR (jmartin@uada.edu); [3]Soil Fertility Specialist, University of Arkansas System Division of Agriculture, Cooperative Extension Service, Little Rock, AR (bfinch@uada.edu); [4]Forages Instructor, University of Arkansas System Division of Agriculture, Cooperative Extension Service, Little Rock, AR (ksimon@uada.edu); [5]Associate Dean and Professor of Animal Science, Arkansas State University, Jonesboro, AR (dkennedy@astate.edu).

GHG EMISSIONS FROM CATTLE EXCRETA IN CONTRASTING GRAZING SYSTEMS

K. Roger Trumpp, J. C.B. Dubeux Jr., J. Portugez Acuña, K. Oduor, I. Lima Bretas, L. Queiroz, R. Sartori Reis, M. Araujo Bernardini, M. Ruiz-Moreno, L. Garcia[1]

Grasslands are used worldwide as the main source of ruminant production with no or low nutrient inputs. Farmers usually add nitrogen (N) through fertilizers or by adding legumes since they can fix atmospheric N by biological N-fixation (BNF). This study compared greenhouse gas emissions from an N-fertilized grass (Grass+N) vs. a grass-legume mixture (Grass+RP) during the warm season. The N fertilized system included an Argentine bahiagrass (*Paspalum notatum* Flüggé) during the warm season, overseeded with winter grasses (oats -*Avena sativa* L.-, and annual ryegrass -*Lolium multiflorum*-) during the cool season (total of 224 kg/N/ha/yr). The grass-legume mixture system consisted of Argentine bahiagrass-Rhizoma perennial peanut (*Arachis glabrata* Benth) during the summer with no N fertilizer overseeded with a mixture of cool-season grasses and a blend of clovers (crimson -*Trifolium incarnatum* L.-, red -*Trifolium pratense* L.-, and ball -*Trifolium nigrescens* Viv.-) receiving 34 kg N/ha/yr. Nitrous oxide emissions were greater for Grass+N compared to Grass+RP from dung and urine, showing that nitrous oxide emissions can be decreased by replacing nitrogen fertilization for legume species. Methane emissions were greater for Grass+RP compared to Grass+N. Carbon dioxide was not different. Including legume species helps reduce greenhouse gas emissions.

[1]MS. Student (ktrumpp@ufl.edu); Professor Forage Agronomy (dubeux@ufl.edu); MS. Student (portuguez.javier@ufl.edu); PhD Student (kenneth.oduor@ufl.edu); Post-doctoral associate (ig.limabretas@ufl.edu); PhD Student (ldantasqueiroz@ufl.edu); Intern (r.sartorireis@ufl.edu); MS Student (m.araujobernardi@ufl.edu); Biological Scientist (ruizm001@ufl.edu); Research Coordinator (lizagarcia@ufl.edu) - North Florida Research and Education Center, Marianna, FL.

IMPLICATIONS OF FLORIDA NEW BIOSOLIDS REGULATION ON BAHIAGRASS PRODUCTION AND NUTRITIVE VALUE

L.O. Vieira-Filho, M.L. Silveira, M.M. Kohmann, J.M.D. Sanchez, E.C. Ricken, A.S. Cardoso, and L.S. Silva[1]

Land application of biosolids (the by-product of wastewater treatment process) to pastures has numerous long-term agronomic and environmental benefits. However, more stringent regulations recently implemented in Florida restrict or, in some circumstances, ban biosolids land application. A 3-year field experiment (2020-2022) was conducted to evaluate the agronomic impacts of reduced biosolids rates imposed by new state regulations. Treatments consisted of 2 P sources (biosolids and inorganic fertilizer) applied at 0, 40, 80, and 120 lb P_2O_5/A/year) applied to an established bahiagrass (*Paspalum notatum* Flueggé) stand. Forage was harvested at 6-week intervals from June to November to determine herbage accumulation, nutritive value, and tissue mineral concentrations. Bahiagrass herbage accumulation increased with P (and associated N) rates, but there was no difference between biosolids and inorganic fertilizer. Biosolids generally resulted in either similar or greater bahiagrass crude protein, in vitro digestible organic matter, and tissue mineral concentrations than inorganic fertilizer. Data demonstrated that biosolids is a viable nutrient source to replace inorganic fertilizer in forage production systems.; however, reduced biosolids rates imposed by new biosolids regulations in Florida will detrimentally impact bahiagrass production and nutritive value.

[1]Postdoctoral Fellow, Clemson University, Edisto Research and Education Center, Blackville, SC (lvieira@clemson.edu); Professor, University of Florida, Range Cattle Research and Education Center, Ona, FL (mlas@ufl.edu); Assistant Professor and Extension Specialist, Department of Agronomy, University of Wisconsin, Madison, WI (mourakohmann@wisc.edu); Postdoctoral Associate, Range Cattle Research and Education Center, University of Florida, Ona, FL (joaomdsanchez@hotmail.com); Master's Student, Horticultural Sciences Department, University of Florida, Gainesville, FL (e.ricken@ufl.edu); Postdoctoral Associate, University of Florida, Range Cattle Research and Education Center, Ona, FL (adasilvacardoso@ufl.edu); Assistant Professor and Extension Specialist, Edisto Research and Education Center, Blackville, SC (lseveri@clemson.edu).

GRAZING MANAGEMENT ON COVER CROPS – HOW INTENSITY CAN AFFECT PRODUCTIVE AND ENVIRONMENTAL OUTCOMES

A. Zubieta[1], M. Wallau[2], Jose Dubeux[3], Emma Matcham[4] and Luana Dantas[5]

Grazing potentializes the benefits of cover crops (CC) in agricultural lands, but productive and environmental outcomes are dependent on grazing intensity. Grazing can maintain on vegetative phase for longer, resulting in greater total biomass accumulation. Grazing can also improve nutrient cycling and reduce nitrate leaching in half as compared to non-grazed systems. Heavy grazing, however, can result in soil compaction in the superficial layer, loss of nutrients (decoupling C and N cycles) and organic matter, and result in reduced individual animal performance. Light-to-moderate grazing intensities increases by 2.8 t/ha or 60% the aboveground litter mass which represents 22 kg N/ha. Through those mechanisms, moderate grazing can reduce in 55% the greenhouse gases emission intensity compared to intensive. Furthermore, actively considering the pasture phase on the fertilization scheme (i.e., systemic fertilization) can improve use of resources, increase total forage production (36%) and crop yield (13%). In all, grazing cover crops can represent great improvements for the agronomic performance and ecological outcome of crop systems, while diversifying and increasing income. For these benefits to occur, it is important to set sound grazing management targets, namely light-to-moderate grazing intensities, via controlling sward height, pre- and post-grazing herbage mass, stocking rate, regardless of the stocking method.

[1]Post-doc Nutrient Management, Agronomy Department, University of Florida, Gainesville, FL (angel.sanchezzub@ufl.edu); [2]Forage Extension Specialist, Agronomy Department, University of Florida Gainesville, FL (mwallau@ufl.edu); [3]Forage Agronomist, North Florida Research and Education Center, Marianna, FL (dubeux@ufl.edu); [4]Assistant professor, Agronomy Department, University of Florida, Gainesville, FL (eg.matcham@ufl.edu); [5]PhD student, North Florida Research and Education Center, Marianna, FL (ldantasqueiroz@ufl.edu).

SCREENING BERMUDAGRASS CULTIVARS FOR BERMUDAGRASS STEM MAGGOT TOLERANCE

Austin J. Wiggins[1], Justin C. Burt[2], Lisa L. Baxter[3], William F. Anderson[4]

Bermudagrass stem maggot (BSM; *Atherigona reversura* Villeneuve) causes significant damage to bermudagrass (*Cynodon* spp.) in southeastern USA. All cultivars are susceptible, but BSM has greater impact on *C. dactylon* cultivars than *C. nlemfuensis* influenced cultivars. Breeding efforts now focus on selection of BSM tolerant genotypes. Sixteen genotypes were selected from the *Cynodon* core collection in Tifton, GA for screening of BSM tolerance. The objective was to compare the agronomic impact of the BSM on 16 experimental genotypes and 5 cultivars that range in BSM tolerance. Paired plots where one plot was sprayed weekly with insecticide and the other served as a non-treated control. Data indicate an interaction of spray × genotype impacted herbage accumulation. Five experimentals generated comparable herbage accumulation to Tifton 85, the known positive standard. Alternatively, there were 8 genotypes in the non-treated controls that had comparable herbage accumulation to Tifton 85. Genotypes with *C. nlemfuensis* influence had greater total digestible nutrients compared to the *C. dactylon* genotypes. Total digestible nutrients were greater and crude protein was less in plots receiving insecticide applications compared to the non-treated controls. Seven genotypes were selected for further evaluation based on agronomic characteristics to reduce resistant development in BSM to pyrethroid.

[1]Graduate Student, Animal and Dairy Science Department (wiggins.austinj@uga.edu); University of Georgia Tifton Campus, Tifton, GA. [2]Postdoctoral Researcher, Crop and Soil Science Department (justin.burt@uga.edu); University of Georgia Tifton Campus, Tifton, GA. [3]Assistant Professor and Extension Forage Specialist, Crop and Soil Science Department (baxterl@uga.edu); University of Georgia Tifton Campus, Tifton, GA. [4]Research Geneticist (bill.anderson@usda.gov), USDA-ARS, Tifton, GA.

ADAPTATION AND PERSISTANCE OF PASTURE AND HAYLAND SPECIES IN A SIMULATED MECHANICAL AND RUMINANT HARVEST IN APPALACHIA

J.I. Wolford, R.K. Lester, and W.S. Haynes[1]

Overharvesting and inadequate fertility management of pastures occur regularly in pastures and hayland in Appalachia and are associated with soil erosion, degraded water quality, limited forage availability, decreased animal performance, animal health issues, and soil health degradation. The objective of this study was to evaluate the effects of harvest management and fertility on select pasture species common to the Appalachian region. Four species/mixes were utilized in the study that were selected based on dominant forage species used in pastures and haylands in the Appalachian region. Experimental treatments were overgrazed/low cutting height and optimally grazed/high cutting height on pastures/hayfields that had added fertility compared to ones with no added fertility. The data after three years suggest that there is higher biomass production potential but a reduction in desired species composition in overgrazed/low cutting height treatments; and lower biomass production potential with maintenance of desired species composition in the optimal grazing/higher cutting height treatments.

[1]Manager/Agronomist, USDA-NRCS, Appalachian Plant Materials Center, Alderson, WV (isaac.wolford@usda.gov); Study Leader, USDA-NRCS, Appalachian Plant Materials Center, Alderson, WV (randall.lester@usda.gov); Biological Technician, USDA-NRCS, Appalachian Plant Materials Center, Alderson, WV (warren.haynes@usda.gov).

AUTHOR INDEX